Exploding Superstars

Understanding Supernovae and Gamma-Ray Bursts

Alain Mazure and Stéphane Basa

Exploding Superstars

Understanding Supernovae and Gamma-Ray Bursts

Dr Alain Mazure
Director of Research
CNRS
Marseille
France

Dr Stéphane Basa
Researcher
CNRS
Marseille
France

Original French edition: *L'Univers dans tous ses éclats: Que se passe-t-il aux confins du cosmos?*
Published © Editions Dunod, Paris 2007
Ouvrage publié avec le concours du Ministère français chargé de la culture - Centre national du livre
This work has been published with the help of the French Ministère de la Culture - Centre National du Livre

Translator: Bob Mizon, 38 The Vineries, Colehill, Wimborne, Dorset, UK

SPRINGER–PRAXIS BOOKS IN POPULAR ASTRONOMY
SUBJECT *ADVISORY EDITOR*: John Mason, B.Sc., M.Sc., Ph.D.

ISBN 978-0-387-09547-9 Springer Berlin Heidelberg New York

Springer is a part of Springer Science + Business Media (*springer.com*)

Library of Congress Control Number: 2008934908

Cover design: Jim Wilkie
Translation Editor: Dr John W. Mason
Typesetting: BookEns Ltd, Royston, Herts., UK

Printed in Germany on acid-free paper

Contents

List of Illustrations

A selection of color plates will be found in the 16-page color section inserted between pages 116 and 117.

Preface

If, as in science fiction films, sound could defy the laws of physics and travel through a vacuum, our universe would reverberate incessantly with the sounds of titanic explosions, coming from its furthest reaches.

The great majority of these explosions arise from the death throes of very massive stars, involving releases of energy so huge that they are among the most energetic events to occur since the very formation of our universe. Such supernovae can liberate, in less than a second, the energy equivalent to that produced by an entire galaxy containing 200 billion stars.

In addition, spy satellites of the Cold War era made astonishing and quite unexpected observations of a new beast in this amazing cosmic menagerie: the *gamma-ray burst*. Unlike supernovae which may shine for days, weeks or even months in the sky, these bursts reveal themselves to us mainly as intense flashes of highly energetic photons, lasting sometimes for only a fraction of a second. We still await a full explanation of the mechanism of these gamma-ray bursts, but it seems that we are dealing with an extremely violent phenomenon involving the collapse and subsequent explosion of a star at least twenty times more massive than our Sun, leading to the creation of a black hole.

Supernovae, seen since the dawn of humanity, and gamma-ray bursts, known to us only during the last forty years, are highly active areas of current cosmic research. In fact, going beyond the quest to comprehend the underlying mechanisms involved in these phenomena, we can state that they have recently become very special tools for cosmologists wishing to undertake detailed studies to further their understanding of the origin, evolution, and composition of the universe in which we live.

These cosmic 'beacons' are used, for example, as 'standard candles' which allow us, like simple surveyors, to make measurements of the very distant universe. We can look back through cosmic time across more than 90 per cent of the age of the universe. So, for example, observations of certain supernovae have revealed for the very first time that around 70 per cent of the energy-matter content of our universe is made up of 'dark energy', the nature of which is as yet completely unknown to us.

Such beacons have also become important 'skymarks', illuminating their immediate surroundings and allowing us to study the cosmos between them and us, like searchlights revealing the matter comprising our universe.

This book will try to throw light on this assemblage of facts, hypotheses and cosmological conclusions, which together enable us to understand the amazing destinies of exploding superstars past and present.

'Among that which is scattered around at random, the fairest thing is the universe'.
'The hidden harmony is better than the obvious.'
'Nature is wont to hide herself.'

Heraclitus of Ephesus

1 Appetizer

'The discovery of a new dish does more for human happiness than the discovery of a new star'[1]

A. Brillat-Savarin

Titanic events happen in the cosmos, on scales unimaginable to the human mind. Some of these events involve releases of energy unequalled since the very formation of our universe in the Big Bang – equivalent to the total energy output of our Sun over its entire 10-billion-year lifetime.

Apart from the interest these explosive events hold in their own right, and in particular the enigma of their origin, their stories are exceptional in several ways. On the one hand, they represent, in their own 'lifetimes', the ultimate stage of cosmic stellar evolution. On the other, they have recently become indispensable tools used by astronomers to plumb the furthest depths of the universe, providing answers to questions about its formation, evolution and composition. Even though these phenomena are linked to individual stars, cosmologists can hardly fail to be interested in them. These exceptional objects, whose lives we shall explore in detail in this book, are the *supernovae* and their far more energetic siblings, *gamma-ray bursts (GRBs).*

Super novae

The term *nova*, introduced by the great pre-telescopic Danish astronomer Tycho Brahe, and signifying 'new' in Latin, was and still is used in astrophysics to designate a category of objects whose brightness increases abruptly. Because these stars suddenly appear in parts of the sky where they were previously unseen, they are called 'novae', since they appear to be 'new stars'.

As will be described later, it was the astronomer Fritz Zwicky who added, early in the twentieth century, the prefix 'super-' to characterize a type of star showing an even more spectacular outburst in brightness. Thus, the term 'supernova' was born, even though the phenomenon had been known to humans for many centuries.

---

1. Even if they disagree with this sentiment, astronomers may still be gastronomers.

Figure 1.1 The expanding shell of debris from the supernova which was seen to explode in 1006. This image is a composite of visible (or optical), radio, and X-ray data of the full shell of the supernova remnant. (NASA, ESA, and Z. Levay (STScI).) See also PLATE 1 in the color section.

Rare – but noticed

Certain brilliant supernovae have not passed unnoticed in the course of human history. Table 1.1 lists the brightest supernovae which have been observed during the past 2000 years.

The supernova of 1006 was probably the most remarkable of them all. Observed by astronomers in Southern Europe, North Africa, the Middle East, China and Japan, it was described as rivaling the half moon in brightness (mag. -9) and was, for a time, bright enough to cast shadows at night. The expanding debris cloud from this incredible outburst, found in the constellation of Lupus, the Wolf, has been imaged at X-ray, visible and radio wavelengths (Figure 1.1).

Table 1.1 Bright Historically Recorded Supernovae

185 AD – Thought to be the earliest recorded historical supernova. Chinese astronomers noted the appearance of a new star in the Nanmen asterism – part of the sky identified with Alpha and Beta Centauri on modern star charts. The new star peaked at about mag. –2 and faded over eight months. Data from two orbiting X-ray telescopes, XMM-Newton and Chandra, indicate that the supernova remnant RCW 86 is the debris from the 185 AD stellar explosion.

1006 – The brightest recorded historical supernova, seen from Southern Europe, North Africa, the Middle East, China and Japan. It peaked at about mag. –9 in early May 1006 in the constellation of Lupus, and probably took at least two years to fade from view.

1054 – Observed from the Middle East, China, Japan and possibly North America, this was the second brightest historical supernova, peaking at about mag. –6 in July 1054, in the constellation of Taurus. It was observable in broad daylight for three weeks and at night for 21 months after outburst. This supernova is the origin of the Crab Nebula.

1181 – Seen by Chinese and Japanese astronomers in the constellation of Cassiopeia. It peaked at about mag. –1 in August 1181, and was visible in the night sky for about six months. The outburst is associated with a radio and X-ray pulsar and the supernova remnant 3C 58.

1572 – Tycho's supernova in Cassiopeia (Figure 1.3). Tycho Brahe described his observations of it in his book *De Nova Stella*. It was this event that steered him towards a career as an astronomer. It peaked at about mag. –4 in November 1572, and took some 15 months to fade from view.

1604 – Kepler's supernova in Ophiuchus. The outburst was observed and documented by Johannes Kepler, although not discovered by him. It was the last supernova to be definitely observed in the Milky Way. It peaked at about mag. –3 in October 1604, and faded over 18 months. Its appearance was used to argue in favour of the Copernican revolution, contradicting Aristotle's idea of an unchanging cosmos.

1987 – Supernova observed within the Large Magellanic Cloud. Known as SN 1987A, the progenitor of the supernova, Sanduleak –69° 202, a blue supergiant star, was identified from pictures taken earlier. This was the brightest supernova to be observed since the invention of the telescope.

Note: We do know of other young supernova remnants in our Galaxy, less than about 2000 years old, where the exploding stars themselves were not definitely seen visually. One example is the bright remnant Cas A (Figure 1.5), which is about 340 years old. The remnant of the most recent supernova in our Galaxy, about 150 years ago at most, has recently been identified from radio and X-ray observations.

This SN 1006 supernova remnant is about 60 light-years across, and is understood to represent the remains of a white dwarf star destroyed in a thermonuclear explosion.

The supernova of 1054 was also very brilliant. It was observed from the Middle East, China, Japan, and Native American pictograms discovered in New Mexico have been interpreted as indicating that it may also have been seen from North America. It is rather surprising that there exists no European record of its

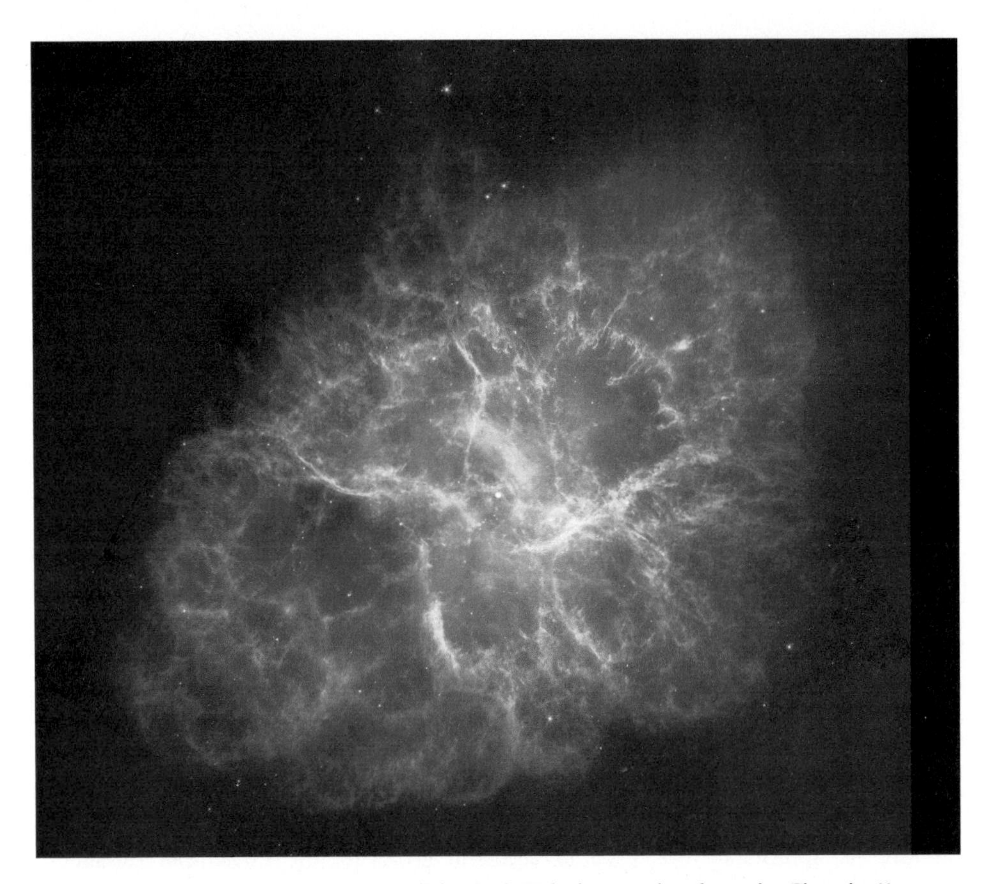

Figure 1.2 This composite image of the Crab Nebula uses data from the Chandra X-ray Observatory, Hubble Space Telescope, and the Spitzer Space Telescope. The central neutron star – the remains of the star which was seen to explode in 1054 – is the bright white dot at the center of the image. (NASA, ESA, CXC, JPL-Caltech, J. Hester and A. Loll (Arizona State Univ.), R. Gehrz (Univ. Minn.), and STScI.) See also PLATE 2 in the color section.

appearance.[2] This supernova peaked at about mag. –6 (brighter than Venus at its most brilliant) in early July 1054, in the constellation of Taurus, the Bull. It was visible in broad daylight for 23 days and at night for over 21 months after outburst. It is this supernova, or rather the resultant expanding cloud of debris, which we may now admire as the Crab Nebula (Figure 1.2). At the heart of this nebula lies a rapidly rotating neutron star - a pulsar - which is the superdense remnant of the massive star that exploded.

2. This phenomenon appears exactly during the great schism between the Church of the West (Catholic) and the Church of the East (Orthodox). Such a coincidence might have been interpreted as bad omen by the authorities and removed from the official notes.

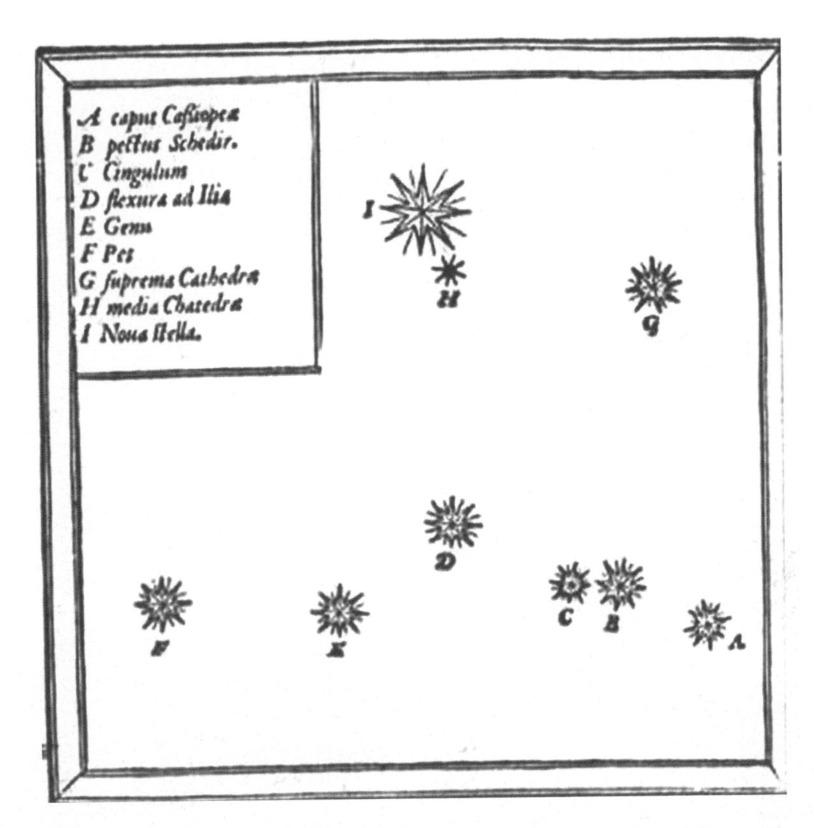

Figure 1.3 When the great Danish astronomer Tycho Brahe was on his way home on 11 November 1572, he noticed a brilliant 'new' star in the constellation of Cassiopeia. This map from Tycho's book *Stella Nova* shows the position of the star among above the stars comprising the familiar 'W' shape of Cassiopeia. (Danish National Library of Science and Medicine.)

The supernovae seen in 1572 and 1604, although not as brilliant as those of 1006 and 1054, were both well documented thanks to the efforts of Tycho Brahe and Johannes Kepler (Figure 1.4), respectively, and their brightness variations were followed as the stars faded from view over many months following the initial outburst. It is indeed unfortunate that both of these events took place only a relatively short time before the invention of the telescope! Their remnants have been studied at X-ray, visible and radio wavelengths, and both events are thought to be the result of white dwarf stars destroyed by thermonuclear explosions. In the case of the 1572 supernova, astronomers may have identified the original companion star of the white dwarf that exploded.

In February 1987, a supernova exploded in the Large Magellanic Cloud (Figure 1.6), a dwarf irregular galaxy which is interacting with the Milky Way. Known as SN 1987A, this was the first (and so far, the only) bright 'modern' supernova which could be observed with large telescopes. Moreover, it was possible to

Figure 1.4 Portrait of Johannes Kepler and of his book *De Stella Nova,* in which he describes his observations of the 1604 supernova. The book includes a map showing the location of the 'new' star. (Harvard-Smithsonian Center for Astrophysics.)

identify the massive, blue supergiant star which had undergone this cataclysm, by examining images of that part of the sky which had been taken previously. This was the first time that such an identification had been possible, and this happy chance led to considerable advances in our understanding of these

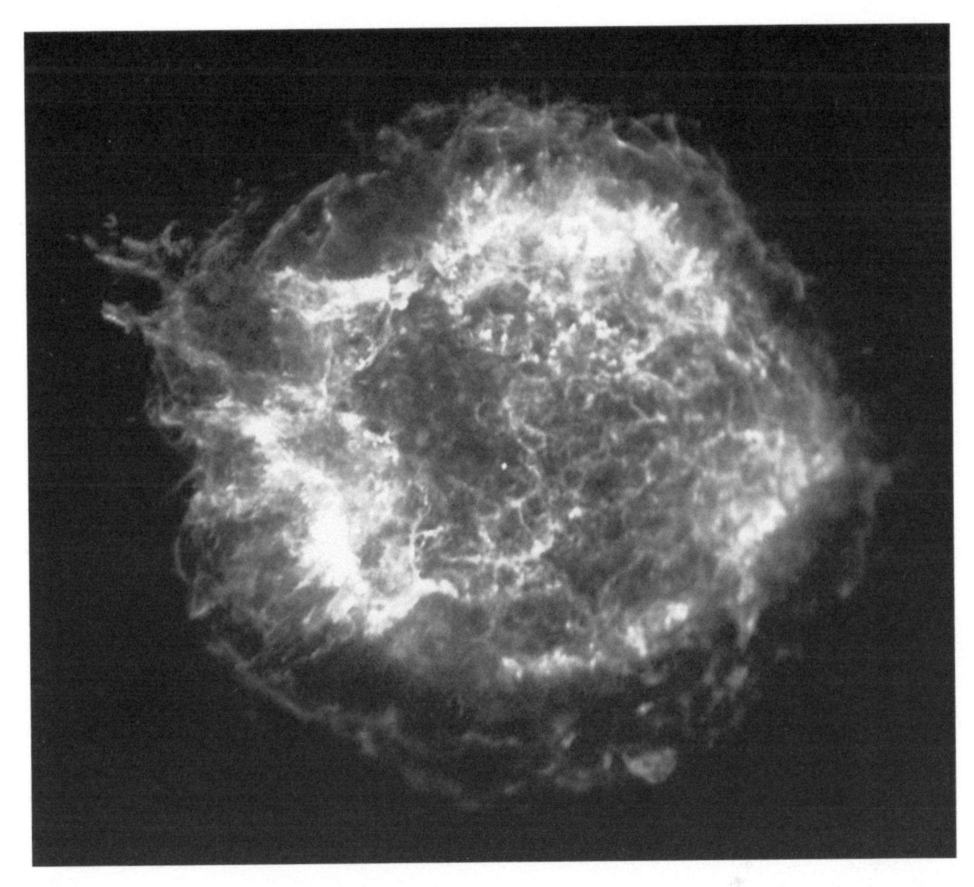

Figure 1.5 Chandra X-ray Observatory image of Cas A, one of the youngest supernova remnants in the Galaxy. A 'hot point-like source' close to the center of the nebula is quite likely the neutron star formed in the explosion of the original star. (NASA/CXC/MIT/UMass Amherst/M.D.Stage et al.)

objects. Unfortunately, this supernova very soon revealed itself to be rather atypical: it had none of the characteristics predicted by theory (its luminosity remained low and the evolution of its light curve over time was unusual). Sadly, this is something so often met in astronomy: there is a general model, but then there are all the special cases!

From stars to cosmology

Even though we still do not fully understand in detail the mechanisms of these cataclysmic events, a certain class of supernova, Type Ia, has captured the attention of astronomers. These supernovae all display a characteristic 'light curve' – the graph depicting the evolution of their luminosity as a function of time – after the initial outburst, and the peak luminosity of the light curve appeared to be consistent across all supernovae of that class. Thus Type Ia

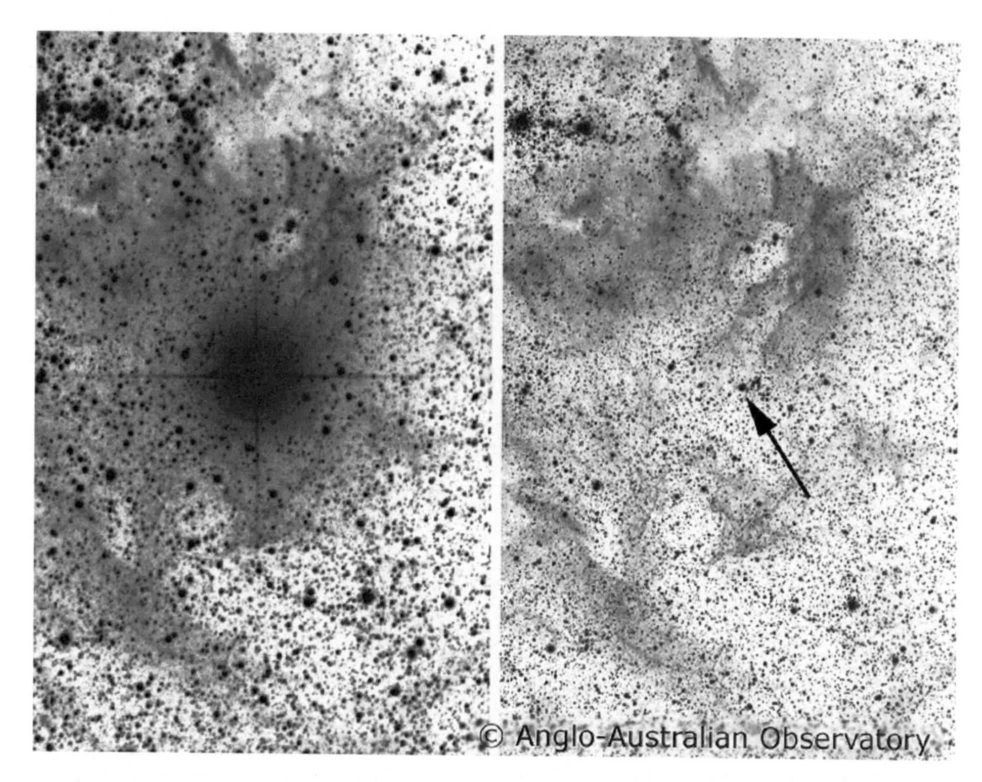

Figure 1.6 Supernova discovered in February 1987 (SN 1987A). As it explodes, the star ejects most of its component matter. This expands more or less isotropically, cooling and forming a nebula (Anglo-Australian Observatory.)

supernovae represented the Holy Grail so long sought by astronomers: they could be used as 'standard candles' (see Chapter 6), allowing astronomers to measure the distances to their host galaxies, and to survey the universe across nearly 8 billion light years, using the information derived to determine its ultimate fate.

Alas, recent discoveries have revealed that things are not quite as simple as this in reality. But, as we shall see, the detailed study of the variations in the brightness of these objects through time, i.e. the analysis of their light curves and of the maxima they display (as seen in Figure 1.7), indicates that they seem to be, if not identical or 'standard', at least capable of being standardized. Thus it has been possible to determine a common procedure for all the objects, such that their light curves are eventually comparable.

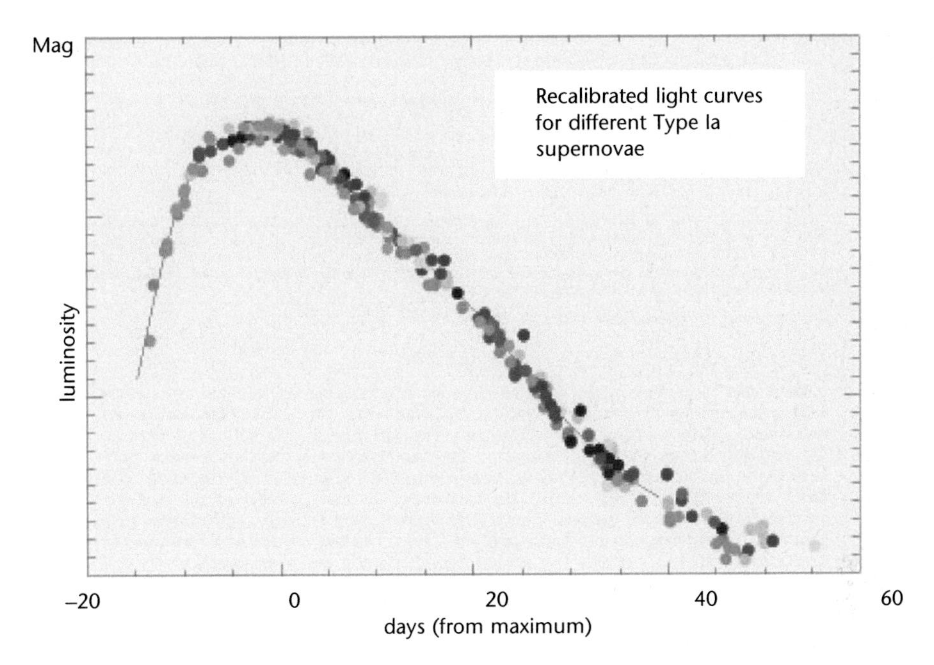

Figure 1.7 Typical shape of the light curve (luminosity as a function of time) of thermonuclear supernovae belonging to one of the two main families of supernovae. The curve is obtained by superimposing curves of the different supernovae after recalibration showing that they are 'standardizable'. The peak luminosity appears to be universal and therefore constitutes a 'standard candle'.

From Earth – or from space?

Celestial flashes

As paradoxical as it may seem at first, the competition between the Americans and the Soviets during the 'Cold War' led to a great number of technical and scientific advances, the greatest of which was, of course, the conquest of our natural satellite, the Moon. Another (understandably, less well known) discovery arose from this incessant rivalry. The story began in 1963, when a nuclear test-ban treaty was signed, involving tests in the Earth's atmosphere. In order to monitor this ban, the US deployed certain military satellites – the Vela satellites (from the Spanish word, Velar, to see) – capable of detecting emissions in the gamma-ray range, which might be evidence of a possible clandestine explosion.

From 1967 onwards, these satellites began to observe flashes of gamma-rays emanating neither from the Earth nor from the Sun. Not until 1973, however, was this major discovery announced to the scientific community (Figure 1.8). The military does not give up its secrets too readily. As Figure 1.9 shows, the very first light curve obtained indicated a short-lived burst of photons emitted at energies consistent with the gamma-ray section of the electromagnetic spectrum.

OBSERVATIONS OF GAMMA-RAY BURSTS OF COSMIC ORIGIN

RAY W. KLEBESADEL, IAN B. STRONG, AND ROY A. OLSON

University of California, Los Alamos Scientific Laboratory, Los Alamos, New Mexico
Received 1973 March 16; revised 1973 April 2

ABSTRACT

Sixteen short bursts of photons in the energy range 0.2–1.5 MeV have been observed between 1969 July and 1972 July using widely separated spacecraft. Burst durations ranged from less than 0.1 s to ~30 s, and time-integrated flux densities from $\sim 10^{-5}$ ergs cm^{-2} to $\sim 2 \times 10^{-4}$ ergs cm^{-2} in the energy range given. Significant time structure within bursts was observed. Directional information eliminates the Earth and Sun as sources.

Subject headings: gamma rays — X-rays — variable stars

I. INTRODUCTION

On several occasions in the past we have searched the records of data from early *Vela* spacecraft for indications of gamma-ray fluxes near the times of appearance of supernovae. These searches proved uniformly fruitless. Specific predictions of gamma-ray emission during the initial stages of the development of supernovae have since been made by Colgate (1968). Also, more recent Vela spacecraft are equipped with much improved instrumentation. This encouraged a more general search, not restricted to specific time periods. The search covered data acquired with almost continuous coverage between 1969 July and 1972 July, yielding records of 16 gamma-ray bursts distributed throughout that period. Search criteria and some characteristics of the bursts are given below.

Figure 1.8 First page of an article from a scientific review, announcing the discovery of gamma-ray bursts of neither terrestrial nor solar origin.

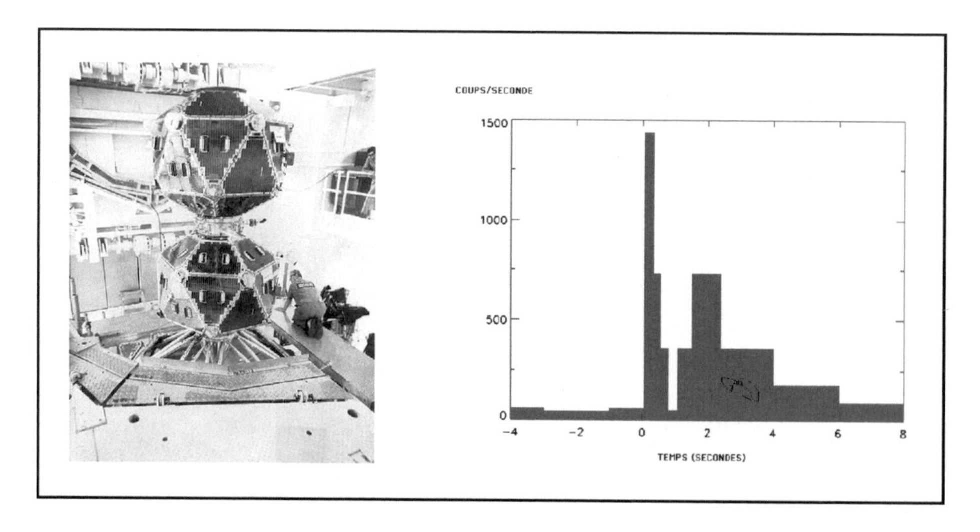

Figure 1.9 Left: a photograph of an American Vela spy satellite. These satellites were mainly used to detect gamma-ray emissions from any violation of the treaty banning atmospheric nuclear tests. Right: the signal of the very first gamma-ray burst observed by the Vela satellite. This phenomenon is characterized by a short-lived burst of photons emitted essentially in the gamma-ray domain.

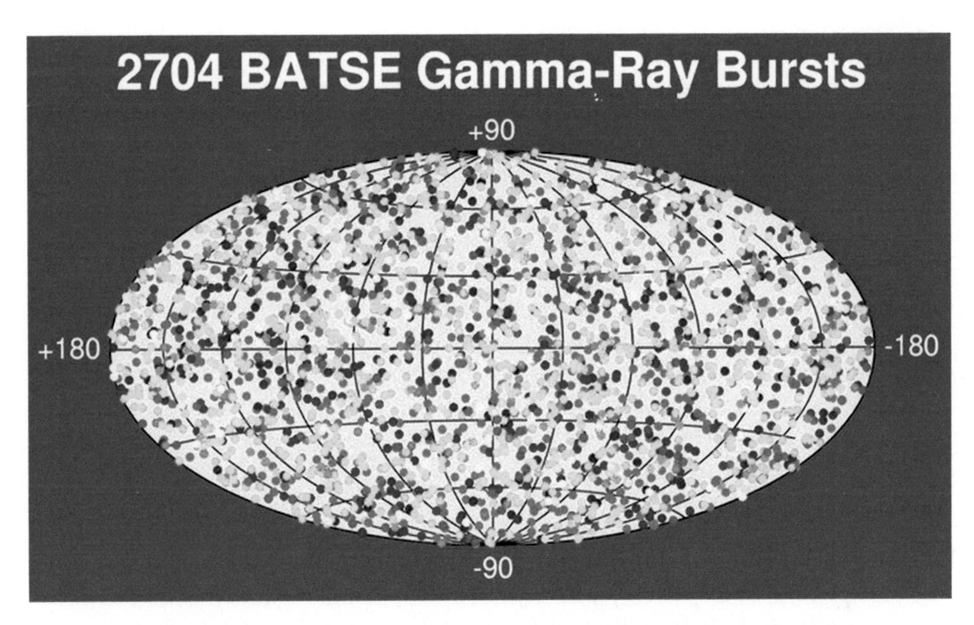

Figure 1.10 Chart showing the positions of 2704 bursts observed by the BATSE instrument on board the Compton Gamma-Ray Observatory. The whole sky is shown, in galactic coordinates (the galactic centre being at 0°/ 0°). Each circle represents an area of a few degrees, corresponding to the uncertainty in the position of the object. The distribution seems largely isotropic. See also PLATE 3 in the color section.

Now, more than thirty years after they were first discovered, several thousand of these gamma-ray bursts have been recorded, mostly thanks to the American Compton Gamma-Ray Observatory (CGRO) and its instrument known as BATSE (Burst And Transient Source Experiment), whose results we see in Figure 1.10. Other important contributors have been the Italian-Dutch satellite BeppoSAX, and more recently the US Swift spacecraft. These new-generation instruments, which first appeared during the 1990s, led to a major breakthrough in our studies of these peculiar celestial objects.

Nowadays, the characteristics of gamma-ray bursts are well known. We shall describe them in greater detail in later chapters. For the moment all we need to know is that the duration of the emissions varies from a fraction of a second to a few minutes, and that the distribution of these durations exhibits two peaks, probably corresponding to two different kinds of families and origins.

The great debate

Echoing the debate in the early twentieth century between Heber D. Curtis and Harlow Shapley concerning the 'extragalactic' nature of nebulae observed in the sky, the question of whether or not gamma-ray bursts were 'local' was long discussed. Several hypotheses gained support within the scientific community. Some thought that the bursts originated from Solar System objects, while others suggested that they occurred inside our Galaxy, either in its plane or in a uniform distribution within the halo surrounding it. Then there were those who argued for an extragalactic origin.

The stakes were considerable: the further away such objects were, the greater the energies which must be involved in order to explain the observations. The advocates of a cosmological origin were therefore very few in number, since, if these events were occurring at the distances they suggested, gamma-ray bursts would represent the most energetic objects observed since the formation of our universe.

A definitive answer to this question came when, in 1997, BeppoSAX observed a burst at wavelengths outside the gamma-ray range – in the X-ray domain: a cascade of observations led to a very accurate measurement of its distance. The finding was unequivocal. The spectral shift of this burst confirmed in no uncertain manner that it was indeed at a 'cosmological' distance: about eight billion light years from Earth.

Since that time, more than a hundred measurements have confirmed this result. The current distance record is held by a burst detected by NASA's Swift satellite on 13 September 2008 (Figure 1.11). The object in question is one of the oldest ever observed: the universe was only about 800 million years old when this burst occurred. At the other end of the telescope, we calculate that the light has taken about 13 billion years to reach us, inviting reflections on the relative scales of time and space.

All the same, and all different

As a final note to this description, we can point out that the light curves of the bursts show very varied rates of evolution, as well as very different durations. Rapid variabilities of the order of a millisecond have been observed, which sets severe limits upon the physical size of the source (approximately 100 kilometers, the distance travelled by light in such a period). We must therefore conclude that the objects responsible for these emissions could be extremely compact. As for the observed spectra, they show a similarity in form (in particular at maximum energy), which points to a common physical mechanism in all these objects. Might they therefore somehow be 'standardized'?

Enigmas to solve – tools to wield

Supernovae and gamma-ray bursts, here described in broad outline, seem to present a certain kinship. Both represent a final stage in the life of a star. Even though the energy emitted by a gamma-ray burst is typically a hundred times that of a supernova, these two celestial cousins are (in company with active galactic nuclei and their massive black holes) the most energetic objects yet discovered in the universe. In both cases, the fact that such an enormous amount of energy can be created and emitted within such short timescales leaves the theoreticians still seeking workable hypotheses.

Such huge quantities of energy are, however, a great gift for astrophysicists as

Figure 1.11 This image of GRB 080913, the most distant gamma-ray burst recorded to date, merges the view through Swift's UltraViolet and Optical Telescope, which shows bright stars, and its X-ray Telescope, which captures the burst, visible near the center of the image. (NASA/Swift/Stefan Immler.) See also PLATE 4 in the color section.

they seek to explore the cosmos across ever greater realms of time and space, with a view to decoding the history and composition of the universe.

In any case, the results suggest that universal mechanisms may be at play within these bodies, offering the hope that these events, so powerful that they can be detected at the edge of the universe, could perhaps be used as 'standard candles'.

Cosmologists would then have within their grasp the means to elucidate both the very early history of our universe – and its ultimate fate.

2 Expanding universe

'Innovation is not the product of logical thought'

Albert Einstein

A very hot universe

Hubble, Einstein and others...

It was Edwin Hubble who, in company with Vesto Slipher and Milton Humason, demonstrated the recession of the galaxies. He established that, with the exception of the nearest systems such as the Andromeda spiral galaxy, our neighbor in the cosmos, galaxies are moving away from the Milky Way at velocities which increase in proportion to their distance.

This led to the famous 'Hubble Law', relating velocity v and distance d:

$$v = H_0.d$$

where H_0 is Hubble's Constant, expressing the recession of the galaxies (as explained in Figure 2.1). This constant in fact varies with cosmic time. This explains the use of the subscript '*0*' to indicate the present time, as with other cosmological time-dependent parameters.

This discovery was, to say the least, extraordinary. An explanation needed to be found. The recession could have been interpreted by assuming that our Milky Way lay at the centre of some phenomenon affecting all the other galaxies. However, this would have meant adopting an anthropocentric vision, ascribing to humankind a special place in the universe: modern cosmology is based upon the *Cosmological Principle*, an extension of the *Copernican Principle*, which rejects the idea that the observer is in some way in a privileged position. Now, according to this principle, the universe will show the same aspect whatever the position of the observer, and in all directions. Therefore, there can be no 'privileged position', and the observed recession involves all galaxies. In other words, the Milky Way as seen from another galaxy would be receding from all the others, and the observation would show the same result whatever the galaxy chosen. No galaxy is at the 'centre' of this general expansion. In order to interpret this phenomenon, we must therefore suppose that it is not

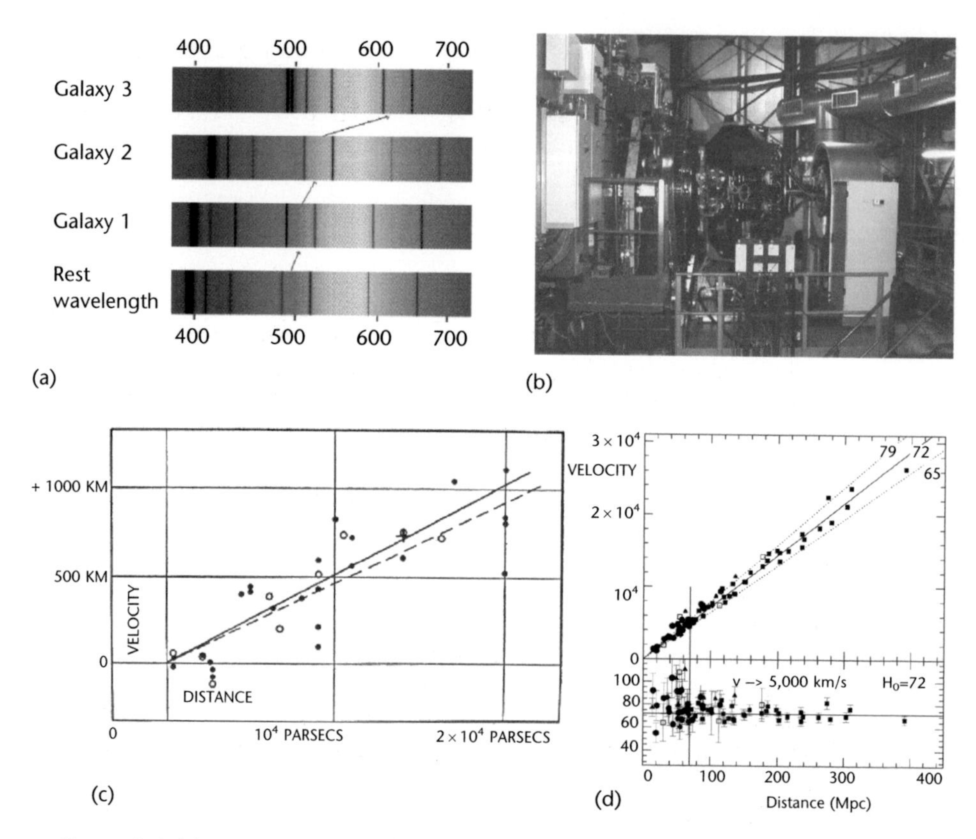

Figure 2.1 (a) Compared with their measured rest positions, the positions of the spectral (emission or absorption) lines (here in nanometers) of a galaxy which is progressively receding from the observer are shifted more and more towards the red. Interpreted as a Doppler effect, this redshift gives a measurement of the velocity of *recession* of the galaxy in question and is proportional to it.

(b) The VIMOS spectrograph, installed on the Very Large Telescope (VLT) in Chile, can measure the velocities of nearly 1000 galaxies in just one exposure. This instrument, the result of collaboration between France and Italy under the aegis of the ESO, takes the form of a 2.5-metre cube and weighs nearly 4 tons.

(c) A graph by Edwin Hubble himself, demonstrating the relationship between the velocity of 'runaway' galaxies and their distance in parsecs (1 parsec = 3.26 light years).

(d) Recent estimates of this relationship, over distances 400 times greater than those of Hubble's era. The result is unequivocal: the further away a galaxy is, the faster it is receding from the observer. The value originally allotted to Hubble's constant of proportionality H_0 was 500 km/s/Mpc (500 kilometers per second per million parsecs from the observer). Current measurements suggest a value of 72 km/s/Mpc. The determination of the constant H_0 was the subject of many controversial debates, engendering numerous revisions. These arose from the difficulty of 'surveying' the universe step by step using different methods of distance estimation according to the scale involved.

the galaxies that move, but rather the space in which they are situated which dilates, or is expanding.[1]

Can we go beyond this intuitive vision, in order to construct a coherent theoretical framework which would allow us to understand this phenomenon and its consequences?

The three pillars of the Big Bang

Let us first return to the postulate of the Cosmological Principle: that the universe is homogeneous and isotropic. What do we observe in reality? We see in fact that this principle applies only generally, i.e. across sufficiently large areas of the cosmos, areas greater than the galaxies and the aggregations that they form. On scales greater than 100 Megaparsecs (Mpc), the distribution of the galaxies is *statistically* similar, whatever their position in space, and in all directions. The most recent and comprehensive observations, as in Figure 2.2, confirm this fact. These properties are also borne out, as we shall see below, by observations of the cosmic background radiation.

It is therefore possible to base a theoretical model on this postulate.

To construct this model in broad outline, let us compare, for a moment, the *recession* of the galaxies to a film, which we can rewind in our minds. Going back in time, we observe the inexorable shrinking of distances and volumes. This continuous condensation implies that densities and temperatures become ever greater, even infinite, as we go further into the past. Such an image is directly identifiable with the basic Big Bang model: a universe, originally very hot and dense, seemingly issuing forth from a 'primordial explosion', or an 'initial singularity'. Ironically, the term 'Big Bang' was coined by Fred Hoyle, a determined opponent of this model: Hoyle's Steady State model, derived from the Perfect Cosmological Principle, was later disproved by observations.

The theoretical framework within which the Big Bang becomes a cosmological model is that of Einstein's General Relativity. In General Relativity, time and space, held to be independent concepts in classical physics, are part of a single four-dimensional continuum, defined through its geometry. The hypotheses of homogeneity and isotropy considerably simplify the equations of General Relativity in its cosmological application. In particular, a universal time can be defined, leading to the establishment of a cosmic chronology.

Einstein also showed that this *geometry* is determined by the *matter-energy content* of space-time. The past, the present and the future of the expansion are therefore determined by the temporal evolution of the matter-energy content of

---

1. It is the expansion of (three-dimensional) space itself that we should envisage, rather than that of the 'content' of that space. The expansion cannot be seen as the result of an explosion whose centre we could observe from without: there is therefore, in the sky, no location of, or direction towards, the Big Bang. The Big Bang occurred everywhere, at the same time (if time has a meaning during this very specific period).

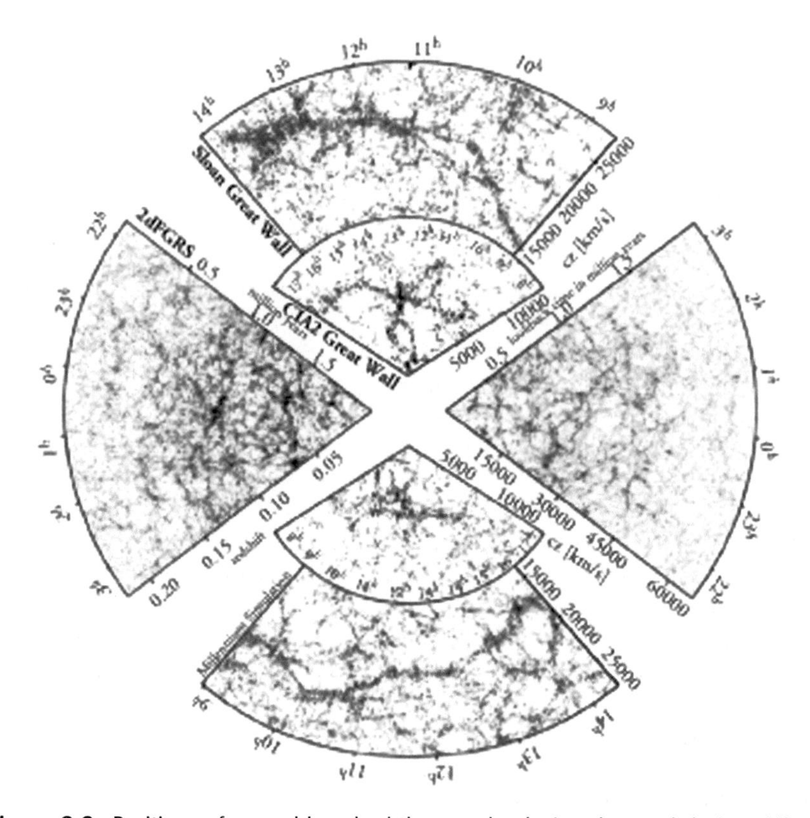

Figure 2.2 Positions of several hundred thousand galaxies observed during different surveys of the cosmos, for example by the Sloan Digital Sky Survey (SDSS) and the 2dFGRS project. These reveal the structure of the universe (top, left and right). Voids and filaments are seen, as well as 'walls' such as the so-called *Great Wall*, extending over tens of Mpc. However, on scales above 100 Mpc, the distribution is similar from one region to another (i.e. statistically homogeneous and isotropic). The two lower diagrams (*Millennium Simulation*) are the result of the largest digital simulation ever undertaken of a tranche of the universe. There is (fair) agreement with reality.

the cosmic 'fluids' filling the universe. The identification at any one moment of the interplay of these cosmic fluids, characterized by their *equations of state*, and the understanding of their evolution, are the key to the thermal history of the universe. Current models allow us to construct this history, from the *Planck time*[2] to the present day, nearly 14 billion years later.

---

2. The Planck time ($t_{Planck} \sim 10^{-43}$ seconds) marks the current limit of our understanding of the beginning of the universe. At this epoch in cosmic time, the universe must be seen as a quantum system, within which quantum mechanics plays as much a part as General Relativity. The theory combining these two aspects is still being worked upon.

As is well known, this model is based upon three 'pillars', whose origins, briefly, lie in:

- the expansion of the universe;
- the formation of the light elements during the primordial nucleosynthesis;
- the cosmic microwave background.

Hubble versus Einstein – but mostly Einstein

Building a cosmological model relies upon the resolution of the equations of General Relativity, wielding the cosmological principle. Thus equipped, we can determine the behavior of a quantity known as the 'scale factor', $R\ (t)$. It is this factor which describes the recession of the galaxies, moving apart as a result of the expansion of the universe. It is measured in practice by the redshift (known as z) of their radiation. From the equations it becomes apparent that a universe dominated by radiation or matter can only be an expanding one.

Now, Einstein, convinced from the beginning that the universe could only be static, introduced into his equations a cosmological constant to counter that inevitable conclusion. The fortunes of this constant have been varied, and we will come upon it in more detail later; suffice it to say for now that Hubble and his co-workers proved Einstein wrong and caused him to recognize his error: thus, the recession of the galaxies became the first pillar of the Big Bang theory.

The other two pillars of this model may be found in the evolution of cosmic energy, which we shall investigate in broad terms. Two immediate and essential elements of this evolution are:

- that matter and energy are equivalent, as described in Einstein's famous relationship: $E = mc^2$;
- that the density and temperature of any cosmic fluid diminish as the universe expands.

So, if the universe, in the beginning, was very hot and dense as we have envisaged above at the start of our little imaginary film, then the energies involved are very high. The primordial cosmic fluid is, in this case, composed of relativistic particles, represented *par excellence* by photons. This period of cosmic history is therefore known as the *Radiation Era.* The temperature gradually fell and, when its value at a given moment, was of the order of that of the rest mass of a given particle, this particle could be created. (Note that, since temperature is falling continuously, the opportunity of mass-energy equivalence for any particle of a given mass m occurs only once.)

We can therefore appreciate that, during the very first seconds and minutes of this scenario, the elementary building bricks of particles (quarks), followed by protons, neutrons and electrons, etc., were formed, to accompany the photons and neutrinos which were already present.

The early battle

Let us dwell for a moment upon the first three minutes in our chronicle. At this time, the particles comprising the cosmic fluid are involved in continuous interactions. Charged particles (protons and electrons) interact electromagnetically with photons; and hadrons (protons and neutrons) with each other (strong interaction). All these particles, and the associated radiation, participate in the inexorable process of cosmic dilution.

As we shall see later, in our investigation into the origin of the energy of stars, if certain physical conditions of temperature and pressure are favorable, nuclear fusion processes between particles can occur. These necessary conditions are in fact fulfilled during the very first minutes of the cosmos, and a veritable alchemy may result. Primordial protons and neutrons tend to join together to form the simplest nuclei found in Mendeleev's periodic table: hydrogen, deuterium, helium, lithium and beryllium... though not without effort, for the repulsive (Coulomb) force between electrically charged particles has to be overcome in order that the strong nuclear interaction, effective only at very small distances, can intervene. Also, expansion itself has to be countered, as particles move apart, discouraging nuclear interaction. Finally – and there is no time to lose – the neutron can exist in isolation for only about fifteen minutes, before becoming a proton.

In this turmoil, in a little less than five minutes, the lightest elements of Mendeleev's table are produced. It is as well that they are, since such an opportunity occurs only once in the history of the universe. The 'winners' in this cosmic competition are, in accordance with the standard cosmological model, hydrogen (about 75 per cent of the mass produced), helium-4 (about 25 per cent), with traces also of deuterium, helium-3, lithium and beryllium; no heavier element is produced at this time. The detailed distribution of elements resulting from this process involves our knowledge of nuclear physics, and, although the calculations are complex, they operate within a well established discipline and (importantly) admit of no 'free' parameter.

The predictions of these abundances are so precise that they constitute a *fundamental test* of the 'hot universe' model and its chronology. To run this test, we must compare the predictions (i.e. the values in the primordial universe aged about 10 minutes) with the abundances observed today (some 14 billion years later). Not an easy task, of course, since the observed abundances have been subject to various physical processes as they evolved. The principle here is to measure them at cosmic sites where there is no such modification of these abundances, or where modification is well understood and models exist for it. Such measurements are difficult to achieve, and are regularly debated and revised.

However, out of this there has arisen an extraordinarily good global agreement between predictions and observations. On this firm basis, the primordial nucleosynthesis of the light elements becomes the second pillar beneath the standard cosmological model.

Let us finish on three important points. Firstly, this test confirms the fact that the laws of physics were already in force about one second after the formation of the universe. This justifies the implicit hypothesis of an intelligible universe where the laws of physics are valid at all times. Secondly, we note that this nucleosynthesis was completed before the production of elements heavier than lithium and beryllium. What then of the rest of Mendeleev's table? The predicted abundances are those of *ordinary 'baryonic' matter*, established for all time across the cosmos. We know therefore precisely, and for any given moment, the quantity of *baryons* in the universe – as we shall see, a not inconsequential fact.

Cosmic fossil

After this episode of nucleosynthesis, the history of the cosmos continued to be one of expansion, dilution and cooling. For a long period, the temperature of the cosmic plasma (and its corresponding energy) was such that the nuclei that had been formed, and the electrons, were still unable to associate to form neutral atoms. However, around 380 000 years after the Big Bang, the temperature had fallen to about 3 000 K, and at that temperature ions and electrons could at last *(re)combine*. This phenomenon, seemingly unremarkable, was in fact an essential and historic occurrence, in more ways than one.

On the one hand, because it was the moment when atoms (at least, the lightest) which constitute our everyday environment, and life itself, were born; and on the other, because this transition from ionized to neutral matter left an indelible *fossil 'signature'*, which astronomers can detect and study. What is the nature of this fossil?

All the while they cohabited, charged particles (ions and electrons) interacted incessantly with particles of electromagnetic radiation (photons), and the result of those interactions was a thermodynamic equilibrium of the whole, the energy distribution of the photons also being in equilibrium and dependent only upon the temperature (in other words, a 'black body'). As recombination occurred, matter became neutral and interactions with photons ceased almost instantaneously. Matter and radiation became *decoupled*.

Immediately, the photons, whose trajectories had been constantly altered in the presence of charged particles, became free to propagate throughout the universe, in all directions. The universe became 'bathed' in photons, all retaining the 'memory' of the temperature (about 3 000 K) which they had at the moment of decoupling. According to this scenario, these photons have come to fill the universe and are present in our environment, still obeying a black-body law at a temperature of about 3 K (the temperature of the decoupling modified by the subsequent expansion). There are now 400 such photons in each cubic centimeter of the universe. It may be of interest to know that some of the 'snow' on our television screens is due to these photons.

This 'photon bath' is one of the predictions of the Big Bang model: its indelible fossil echo. This *cosmic microwave background* (CMB) was detected by chance in the mid-1960s by Arno Penzias and Robert Wilson (Figure 2.3), and its discovery earned them the Nobel Prize in 1978. The American COBE satellite

Figure 2.3 Nobel Prizewinners Arno Penzias (left) and Robert Wilson, discoverers of the cosmic microwave background in the mid-1960s. They are standing on the horn-shaped microwave antenna at Holmdel, New Jersey, with which they made the discovery accidentally, while using the supersensitive antenna to detect the faint radio waves bounced off orbiting balloon satellites. (AIP Emilio Segre Visual Archives, Physics Today Collection.)

established definitively in the 1990s that the CMB is indeed a 'black body' at a temperature of 2.726 K (see Figure 2.4). The determination of its exact value led to another Nobel Prize in 2006, for George Smoot and John Mather.

The Big Bang model rested at last upon its third pillar.

Dark universe

Let us pursue our 'anatomy of the cosmos' by now looking at the modern, 14-billion-year-old universe, and dissecting its energy-matter content.

As far as radiation is concerned, we now appreciate, thanks to many instruments both in space and on the Earth, the various contributions to the electromagnetic spectrum, which is largely dominated by the CMB, the fossil echo of the hot, dense primordial universe. When we consider matter, we have only to look up into a clear dark sky, on a summer night, with binoculars or a small telescope, to discover the myriads of stars which make up that immense pearly-white band known as the Milky Way (Figure 2.5).

The Milky Way is in fact only one unremarkable galaxy among billions populating our universe, as revealed to us, for example, in deep-field images taken by the Hubble Space Telescope (Figure 2.6). These galaxies, spiral, elliptical and irregular in form, constitute the 'building blocks' of an expanding cosmos.

Weighing the universe

If the totality of these 'blocks' contains all the matter in the universe, we can therefore 'weigh' it by estimating the typical masses of its constituent elements, the galaxies. This might not seem a difficult task at first sight, if the mass of each galaxy is simply the sum of the masses of all its individual stars, to which we can add that of interstellar gas and dust. This method will reveal to us the 'luminous mass' of the galaxies.

Similarly, we ought to be able to measure the masses of the gigantic *clusters of galaxies* (Figure 2.7), with their tens or even hundreds of members, by adding together the masses of the individual galaxies within them. However, the detection within these clusters of a hot plasma at a temperature of around 10^8 K, emitting X-ray radiation, has shown that we should also add the contribution of this gas, which in fact very largely dominates the luminous mass of the *ensemble* of the galaxies.

There are other techniques we can use to determine the masses of galaxies and clusters of galaxies. One is based upon the dynamics of these systems, which are deemed to be in equilibrium. The method involves studying the rotation of spiral galaxies on their axes, and the movement of galaxies within clusters. The rate of rotation of the spirals or the motions of galaxies within clusters is a reflection of the total mass of these systems. The application of the fundamental laws of mechanics (see Appendix 1, page 125) allows us easily to relate these quantities, which are measurable, to the *dynamical mass*, i.e. the mass determined by the characteristics of the motions of the objects involved.

To astronomers' considerable surprise, this dynamical mass has been found to be far greater than the luminous mass. This excess, with the dynamical mass outweighing the luminous mass by 5 to 10 times, leads naturally[3] to the notion of *dark matter*.

The deficit in luminous matter is confirmed by observations of gigantic arcs within clusters of galaxies (Figure 2.8), with light from background objects being distorted by the total mass of these systems – in General Relativity, the notion of force gives way to that of the curvature of space in the presence of matter-energy, and the mass of the clusters of galaxies is locally curving space, causing the photons to follow geodesics. This curvature of 'light rays' causes *gravitational lensing* effects, similar to the effects of optical lenses. It is possible, given the observed deflections, to determine independently the total mass of the system in question.

---

3. An alternative possibility, involving the modification of the laws of gravity (as in the Modified Newtonian Dynamics or MOND theory), does not seem to be borne out by the evidence.

(a)

(b)

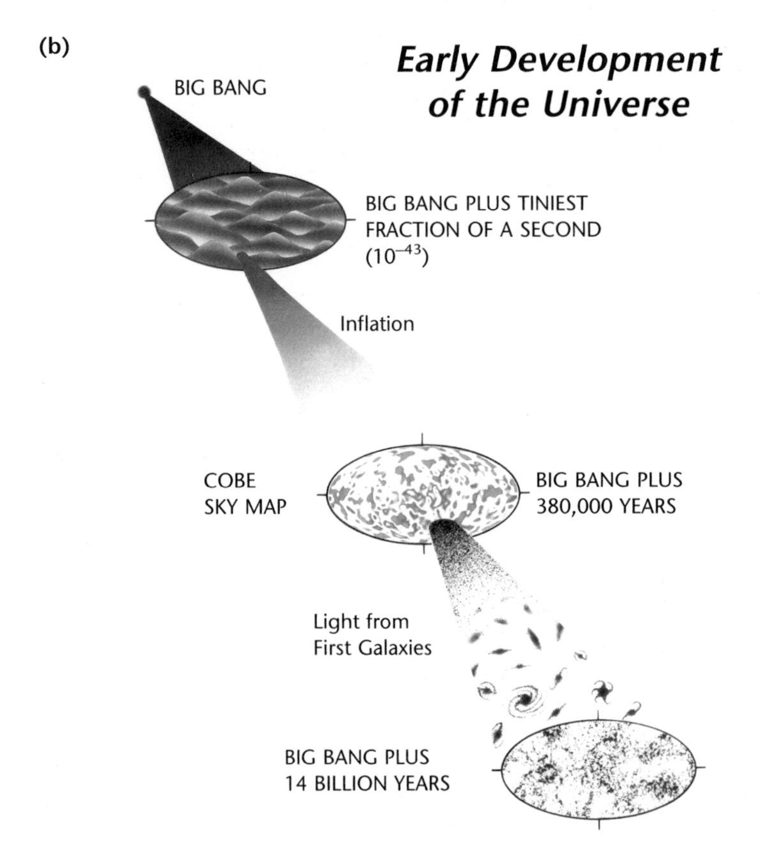
Early Development of the Universe
BIG BANG
BIG BANG PLUS TINIEST FRACTION OF A SECOND (10^{-43})
Inflation
COBE SKY MAP
BIG BANG PLUS 380,000 YEARS
Light from First Galaxies
BIG BANG PLUS 14 BILLION YEARS

(c)

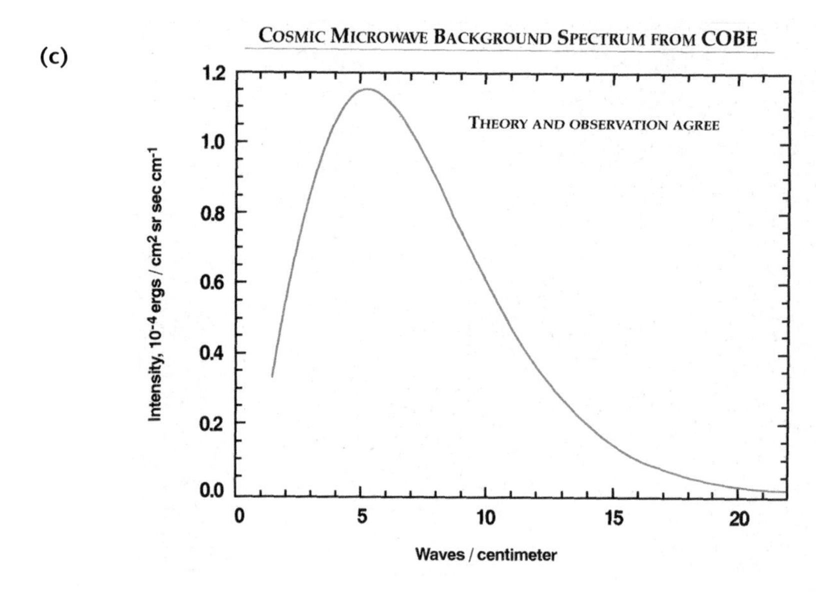

Figure 2.4 (a) An artist's impression of the COBE (COsmic Background Explorer) satellite in Earth orbit. See also PLATE 5(a) in the color section.
(b) Cosmic history since the Big Bang. After the Planck time (about 10^{-43} s), for the epoch beyond which a satisfactory theory unifying gravity and quantum mechanics remains to be elaborated, there occurred an inflationary phase during which the universe expanded exponentially. The primordial fluctuations which gave rise to the galaxies we see today were generated at this epoch. After 380,000 years, photons decoupled from matter and flooded out freely through the universe. They constituted a perfect black body, detected by COBE at a temperature of approximately 2.73 K. See also PLATE 5(b) in the color section.
(c) The intensity of the sky background radiation as measured by COBE as a function of wavelength, exactly matching the predictions for a perfect black body at a temperature of 2.73 K. (NASA Goddard Space Flight Center.)

Since the amount of 'ordinary' matter is insufficient to explain these distortions, this method, and others already mentioned involving the dynamics of galaxies and clusters, necessitates bringing dark matter into the equation. Dark matter is different from ordinary, baryonic matter, and involves massive, neutral particles having almost no interaction with ordinary matter.

In the standard model of particle physics, the *neutrino*[4] fulfils some of these criteria; during the 1980s, it seemed that this might be the 'most wanted' particle, but its lack of mass caused it to fall from the limelight. It could not provide answers to the challenging questions laid down by the dynamics of

4. A particle predicted by the physicist Pauli in the 1930s and discovered experimentally in 1956 (leading to a Nobel Prize in 1995 for the discoverers, Frederick Reines and Clyde Cowan). Very recent experiments assign it a very small, but non-zero mass.

Figure 2.5 The Galaxy looks like a great milky band extending across the whole sky. In the southern hemisphere its two smaller neighboring galaxies can be seen: the Small and Large Magellanic Clouds. (Cerro Tololo Interamerican Observatory.)

galaxies and clusters; nor could it explain the scenarios of the formation of the major structures of the universe.

Fortunately for cosmologists, modern developments in particle physics suggest various possible candidates. Among them, and the favorite, is the *neutralino*, predicted by certain physical ('supersymmetry') models. It possesses the singular ability to be stable and electrically neutral (hence its name, small neutral particle in Italian). Research into this particle and its siblings is intense, since it is a key to our comprehension of the world of particles and, at the same time, the universe.

Decelerated or accelerated?

The recession of the galaxies points to the expansion of the universe, the theoretical framework of which resides in the Big Bang model.

In its standard version, the thermal history of the universe from the Planck time onwards is essentially characterized by two great eras: first, an era during which the energy-matter content of the universe was dominated by radiation (the Radiation-Dominated Era), and second, the Matter-Dominated Era, during which it was the material content that prevailed (see Figure 2.9). We can

Figure 2.6 This million-second-long exposure, called the Hubble Ultra Deep Field (HUDF), reveals the first galaxies to emerge from the so-called 'Dark Ages', the time shortly after the Big Bang when the first stars reheated the cold, dark universe. This view is actually two separate images taken by Hubble's Advanced Camera for Surveys (ACS) and the Near Infrared Camera and Multi-object Spectrometer (NICMOS). Both images reveal galaxies that are too faint to be seen by ground-based telescopes, or even in Hubble's previous faraway looks, called the Hubble Deep Fields (HDFs), taken in 1995 and 1998. (NASA, ESA, S. Beckwith (STScI) and the HUDF Team.) See also PLATE 6 in the color section.

therefore, in theory, easily predict the 'way ahead' for cosmic expansion; the universe now being dominated by matter, we can expect the expansion to slow because of the gravitational effect of that matter – a similar situation to that of the classic case of a projectile gradually slowing under the influence of the Earth's attraction. A great quantity of matter will have a decelerating effect upon anything moving away in its vicinity.

So imagine the surprise of scientists when, in 1998, two groups of researchers using Type Ia supernovae as 'standard candles' (see Chapter 6) showed evidence

Figure 2.7 NASA's Hubble Space Telescope captures the magnificent Coma Cluster of galaxies, one of the densest known galaxy collections in the universe. Hubble's Advanced Camera for Surveys viewed a large portion of the cluster, spanning several million light-years across. The entire cluster contains thousands of galaxies in a spherical shape more than 20 million light-years in diameter. (NASA, ESA, and the Hubble Heritage Team (STScI/AURA).) See also PLATE 7 in the color section.

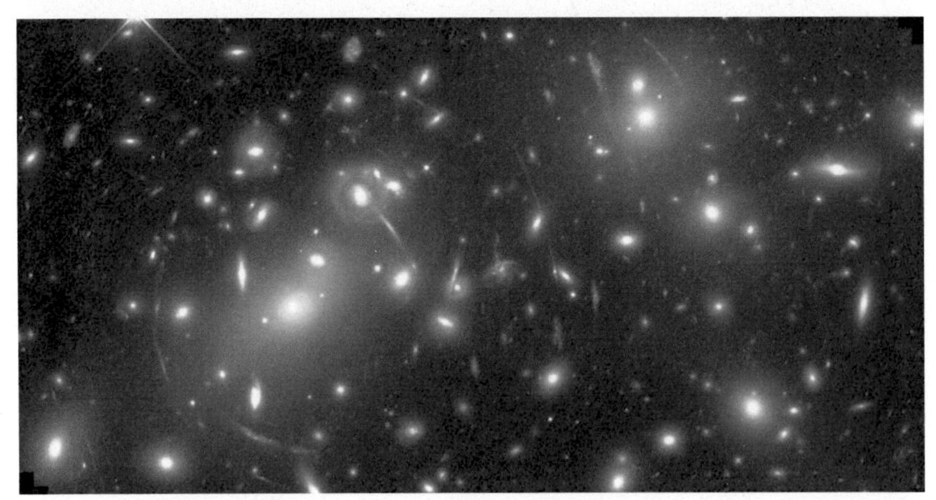

Figure 2.8 This Hubble Space Telescope image of a rich cluster of galaxies called Abell 2218 is a spectacular example of gravitational lensing. The arc-like patterns spread across the picture like a spider's web is an illusion caused by the cluster's gravitational field. This cluster of galaxies is so massive and compact that light rays passing through it are deflected by its enormous gravitational field, much as a camera's lens bends light to form an image. This phenomenon magnifies, brightens, and distorts images of those faraway objects. (Andrew Fruchter (STScI) et al., WFPC2, HST, NASA.) See also PLATE 8 in the color section.

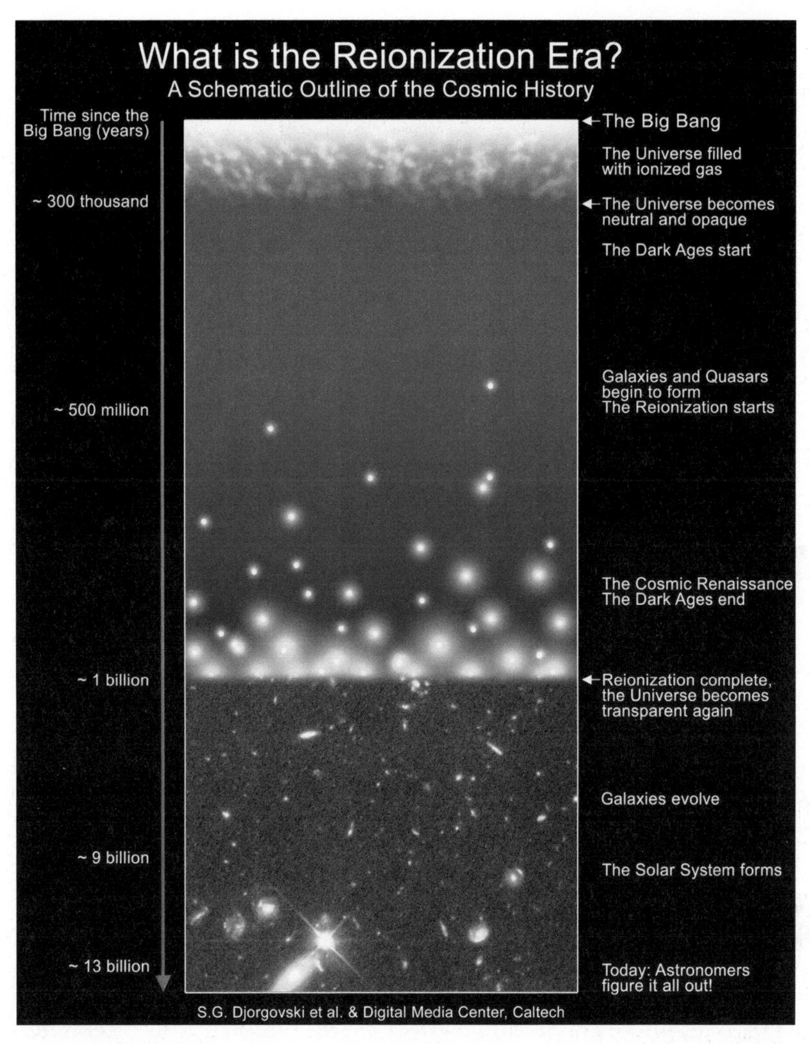

Figure 2.9 The thermal history of the universe. The universe started with the Big Bang nearly 14 billion years ago, and from the Planck time onwards is generally characterized by two great eras. First, there was an era during which the energy-matter content of the universe was dominated by radiation (the Radiation-Dominated Era), and second, the Matter-Dominated Era, during which it was the material content that prevailed. Eventually, by around 380,000 years after the Big Bang, atomic nuclei and electrons had combined to make atoms of neutral gas. The glow of this 'Recombination Era' is now observed as the cosmic microwave background radiation. The universe then entered the 'Dark Ages', which lasted about half a billion years, until they were ended by the formation of the first galaxies and quasars. The light from these new objects turned the opaque gas filling the universe into a transparent state again, by splitting the atoms of hydrogen into free electrons and protons. This Cosmic Renaissance is also referred to by cosmologists as the 'Reionization Era', and it signals the birth of the first galaxies in the early universe. (S. G. Djorgovski et al., Caltech and the Caltech Digital Media Center.)

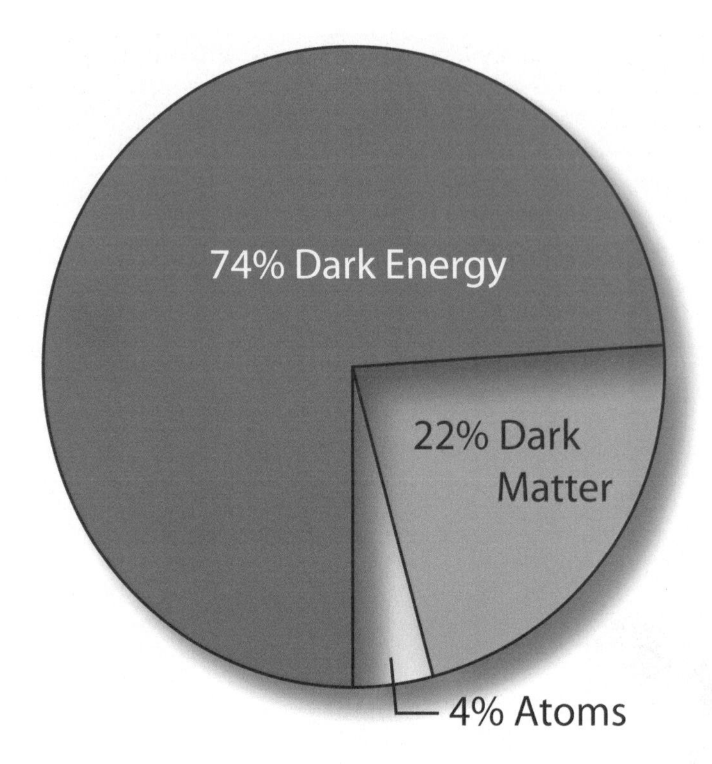

Figure 2.10 The 'cosmic budget'. The universe is dominated by unseen material: dark matter and dark energy. Baryonic ('ordinary') matter (with radiation) makes up only about 4% of the total. (NASA/WMAP Science Team.)

for an *acceleration* of the expansion, instead of the expected deceleration. In order to explain this phenomenon, it had to be accepted that that the evolution of the universe had been, for approximately the last 5 billion years, dominated not by matter but by a 'cosmic fluid' possessing the strange property of exercising what amounts to a 'repulsive gravitational effect'.[5] This period is now often referred to as the Dark-Energy-Dominated Era.

Strange as it may seem, such 'fluids' are predicted by fundamental physics. Unlike 'traditional' fluids, they have an equation of state (see Appendix 2, page 127) within which 'pressure' may be negative. *Vacuum energy* (or one of its incarnations) possesses this surprising property.

Consequently, we are asked seriously to envisage a type of energy created at the time of the earliest universe, and reappearing 9 billion years later as the dominant factor in its destiny.

5. For the purposes of simplification, we use the idea of gravitational force here, even though General Relativity abandoned the concept in favor of the curvature of space.

After this cosmic *tour d'horizon*, and the evaluation of the different components of matter and energy, we can work out a 'balance sheet' (Figure 2.10) of the contents of the universe. Its unexpected conclusion is that the 'visible' universe of planets, stars, galaxies, living things etc., emitting radiation across the range of the electromagnetic spectrum, is but a small fraction (less than 5 per cent) of the total mass of the universe.

So the universe is predominantly dark, dominated by dark energy thought to represent about 74 per cent of the total. The rest is essentially matter, but nearly all of that is also dark.

3 From the universe to the stars

'I shall ascend to infinite space, I shall traverse the spirit of the Earth, I shall journey in light, and I shall reach the star'

Poem of the Egyptian Middle Kingdom.

From quantum clumps to the first light

If the universe has been expanding for almost 14 billion years; if it is dominated by essentially non-baryonic constituents; and if moreover it is homogeneous and isotropic, according to the Cosmological Principle: then *when* and *how* did great inhomogeneities such as stars, galaxies and the large-scale structures of the universe, form?

Instability understood

This question remains among the most crucial in modern astrophysics. The last two decades have seen important progress and, although some points remain to be elucidated, it is thought that the question has been broadly answered.

The phenomenon responsible is based on a mechanism known as 'gravitational instability'. In this scenario, the major structures of the universe (stars, galaxies and clusters of galaxies) are all the result of very small excesses (or *superdensities*) of matter which underwent remarkable growth under the influence of gravity, to become the celestial bodies that we observe today.

As we shall see, this is a mechanism that requires several prerequisites. The first of these is the existence of small density fluctuations in the primordial 'fluid' of matter, which act as 'seeds' for the emergence of large cosmic structures. However, the standard Big Bang model contains no prediction of such primordial fluctuations in the density of matter, and we have to postulate their existence. So we come to the question of the origin of these inhomogeneities in this particular context.

One of the major cosmological theories put forward during the 1980s involved the inclusion into the standard model of a so-called 'inflationary' phase. According to physicists Alan Guth and Andrei Linde, the cosmos underwent a period of exponential expansion not long after the Planck time (Figure 3.1), and under the influence of a type of energy similar to the dark energy encountered in

Figure 3.1 A representation of the evolution of the universe over 13.7 billion years. The far left depicts the earliest moment we can now probe, when a period of 'inflation' produced a burst of exponential growth in the universe. (Size is depicted by the vertical extent of the grid in this graphic.) The afterglow light seen by WMAP was emitted about 380,000 years after inflation and has traversed the universe largely unimpeded since then. The conditions of earlier times are imprinted on this light; it also forms a backlight for later developments of the universe. (NASA/WMAP Science Team.) See also PLATE 9 in the color section.

the previous chapter. This expansion resulted especially in the *spatial flatness* of the universe, as observed in studies of the cosmic microwave background.

One virtue of this inflationary scenario is that it actually predicts the existence (and also the intensity) of initial density fluctuations of quantum origin. This prediction, and also the overall flatness of space, was recently confirmed by measurements taken by the WMAP (Wilkinson Microwave Anisotropy Probe) satellite (Figure 3.2 (a)).[1] These showed that the overall temperature distribution of the cosmological 'black body' was extremely uniform (Figure 3.2 (b)), having an average temperature of 2.725 Kelvin (degrees above absolute zero; equivalent

1. When WMAP observed the microwave background sky it looked back to when there were free electrons that could readily scatter cosmic background radiation. This cosmic background 'surface' is called the 'surface of last scatter'. If there were any density fluctuations imprinted in this surface of last scatter (represented by regions that were very slightly hotter or cooler than average) they will remain imprinted to this day because the emitted radiation travels across the universe largely unimpeded.

Figure 3.2 (a) The Wilkinson Microwave Anisotropy Probe (WMAP) used the Moon to gain velocity for a slingshot to the Lagrange point L2. After three phasing loops around the Earth, WMAP flew just behind the orbit of the Moon, three weeks after launch. Using the Moon's gravity, WMAP stole an infinitesimal amount of the Moon's energy to maneuver into the L2 Lagrange point, 1.5 million km beyond the Earth. (NASA/WMAP Science Team.)

to –270°C.), but with minute temperature variations of just a few ten thousandths of a degree across the sky (Figure 3.3). It can be shown that these temperature fluctuations indeed represent the initial fluctuations in the density field of matter. So here we have, thanks to this new theory, the original quantum 'clumps', and knowledge of their amplitude.

The second prerequisite is a kind of 'growth mechanism' transforming these small inhomogeneities into the major structures of the universe. This mechanism is in fact none other than gravity: if small local 'superdensities' exist in a distribution of matter of mean uniform density, they will grow simply by attracting surrounding matter with their gravity.

If nothing intervenes to halt this process, it can theoretically go on forever. This mechanism, the original model for which we owe to physicist and astronomer James Jeans, is called 'gravitational instability'. Now, matter collapsing upon itself in this way will transform the resultant energy (gravitational potential energy) into kinetic energy. The medium involved will be warmed and pressure will develop within it, engendering forces which can counterbalance the effects of gravity. It can be shown that, as a result of this, there exists a *limiting mass* known as the 'Jeans mass', whereby equilibrium between the two forces is reached and the existence of stable objects becomes possible.

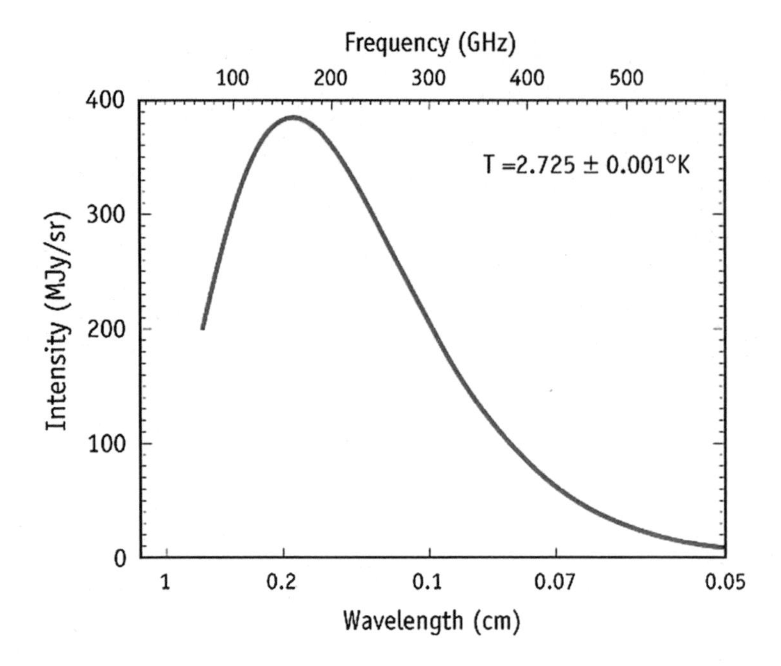

Figure 3.2 (b) This figure shows the prediction of the Big Bang theory for the energy spectrum of the cosmic microwave background radiation compared to the observed energy spectrum. The FIRAS experiment on WMAP measured the spectrum at 34 equally spaced points along the blackbody curve. The error bars on the data points are so small that they can not be seen under the predicted curve in the figure! There is no alternative theory yet proposed that predicts this energy spectrum. The accurate measurement of its shape was another important test of the Big Bang theory. (NASA/WMAP Science Team.)

Stars in the dark

The scenario described above applies to a static environment. In the case of an expanding universe, the growth of initial density perturbations is in reality much less rapid than in the case of a static universe. In fact, the expansion, which tends to dilute the 'fluid' of the matter, is constantly competing with the tendency of the object being formed to collapse upon itself, as a result of its own gravity.

Is it nevertheless possible for the major structures which we observe in an expanding universe to form according to this scenario? The surprising paradox is that the ordinary (baryonic) matter of the universe cannot give rise to the celestial bodies in our universe. So how is it that they populate it, and came into being?

At the beginning of the thermal history of the universe, the growth of the extra-dense baryonic matter, composed of protons and neutrons, was 'frozen' by interactions between these protons and electrons with photons from the cosmic microwave background, to such an extent that the matter remained ionized. The photons, which were much more numerous than the baryons, imposed their

Figure 3.3 The detailed, all-sky picture of the infant universe from three years of WMAP data. The image reveals 13.7 billion year old temperature fluctuations (shown as differences in tint) that correspond to the seeds that grew to become the galaxies. The signal from our Galaxy was subtracted using the multi-frequency data. This image shows a temperature range of $\pm$ 200 microKelvin. (NASA/WMAP Science Team.) See also PLATE 10 in the color section.

particular qualities and robbed the baryons of all independence of action. Also, there was not enough baryonic matter sufficiently to slow the expansion and allow the triggering of the process of instability envisaged by Jeans, leaving the field open for local gravitational action. So the growth of the baryonic 'clumps' was severely hampered.

Dark matter, being neutral, and present in greater quantities, was not subject to these contrary factors and therefore did not interact with photons. It could by its very abundance counteract the expansion. In this model, 'haloes' of dark matter are formed, and within them, ordinary, now neutral matter could condense. So, billions of years ago, galaxies and the primordial intergalactic medium came into being, to become the birthplaces of the first generation of stars. At this epoch of cosmic history, this medium represented only the plasma created during the primordial nucleosynthesis, when the first elements of Mendeleev's table were forged. It consisted of approximately 75 per cent hydrogen and 25 per cent helium, with small percentages of other, heavier elements, as we have already noted.

This medium was by no means perfectly homogeneous, and density fluctuations within it meant that Jeans' mechanism could operate on a smaller scale. Clouds of matter condensed, drawing in ever more material from their surroundings. At the centers of these early aggregations, pressure and temperature rose, establishing an equilibrium after a few hundred thousand years. Matter, heated to high temperatures, was now able to radiate. So, the very first stars were born, thanks to dark matter.

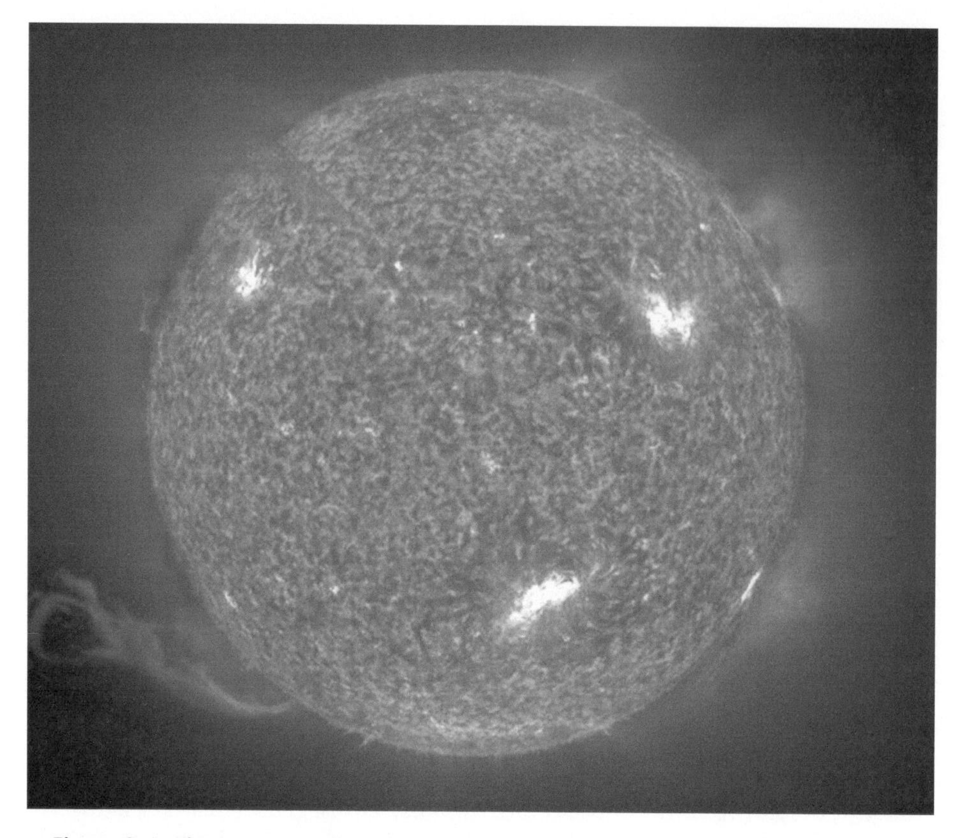

Figure 3.4 This X-ray image of the Sun, taken by the SOHO satellite, shows numerous active regions in the Sun's atmosphere. The hottest and most active regions appear white, and the darker areas indicate cooler temperatures. The wispy feature in the lower left portion of the disk is a solar prominence, a huge cloud of relatively cool plasma suspended in the Sun's hot thin corona. (SOHO (ESA + NASA).) See also PLATE 11 in the color section.

Nuclear stars

One question that taxed the minds of physicists for many years was: where do stars find the energy which allows them to shine for periods as vast as several billion years? Our Sun, for example, continuously emits radiation equivalent to 10^{27} watts – the output of 10^{18} nuclear power stations; that is a billion billion power stations. And it has been doing this for the last 5 billion years (Figure 3.4).

Nineteenth-century physicists tried to explain stellar energy as the product of some classic process of combustion, or as the transformation of potential energy into kinetic energy as the star slowly 'collapsed' in upon itself (see Appendix 5, page 134). In either case, the star's lifetime would not exceed a few million, or tens of millions, of years; such a value was in contradiction with the age of the Earth, which was correctly known at the time. It was left to two physicists, Hans Bethe and Arthur Eddington, to provide the answer, basing their theory on relativity, which established the equivalence of mass and energy, and on contemporary advances in nuclear physics.

In the extreme physical conditions prevailing at the cores of stars[2] such as our Sun, protons can fuse[3] to former heavier and more complex elements. If the mass of the resulting nucleus is smaller than the sum of the masses of the original particles, the difference is transformed into energy, in accordance with the law of the conservation of energy and the famous relationship $E = mc^2$.

Nuclear fusion reactions are at the origin of the energy of stars. Among all the possible reactions, the *proton-proton cycle* is the predominant one inside stars like our Sun (Figure 3.5). Here, four protons form a helium nucleus, together with energy in the form of photons and neutrinos.

So now we are able to answer the question: what is a star? It is a celestial body subject to its own gravity (a 'self-gravitating' system) and of sufficient mass to trigger nuclear fusion reactions within its core.

The masses of the stars

For the energy radiated by stars to be the product of thermonuclear reactions, certain minimum conditions of density and temperature have to be fulfilled. Only gravity, derived from the mass of the star, can create such conditions. Not every self-gravitating body will therefore be the seat of such reactions, as is the case with planets – failed stars. The evolution of a star is regulated by one parameter alone, and that is its mass[4] (Figure 3.6).

Photons under pressure

Photons, like ordinary particles, exert pressure; in the case of photons, it is known as *radiation pressure*. This pressure can act together with 'thermal' pressure (due to the agitation of particles 'heated' to a temperature T) to counteract gravity which might lead to the collapse of the star upon itself. It may happen that this radiation pressure becomes dominant, not only as compared with the thermal pressure, but also to the extent that it overcomes gravity and leads to a veritable evaporation of the star. The temperature at which this occurs is known as the *Eddington limit*, and it corresponds to an upper limit for the mass of stars, of the order of 120 times the mass of the Sun.[5] Beyond this limit, stars would fly

---

2. Core temperatures can exceed 10 million degrees, and densities more than 100 times that of water!
3. Fusion reactions involve at least two nuclei combine to form a more massive nucleus; fission reactions involve a massive nucleus splitting into a number of lighter nuclei.
4. Simple considerations based on fundamental physics (see Appendix 3, page 130) can determine the limits of this 'mass regime'
5. This is the unit of mass conventionally used for stars or other massive objects. The mass of the Sun is around 2×10^{30} kg. It is denoted by the symbol $M_{\odot}$.

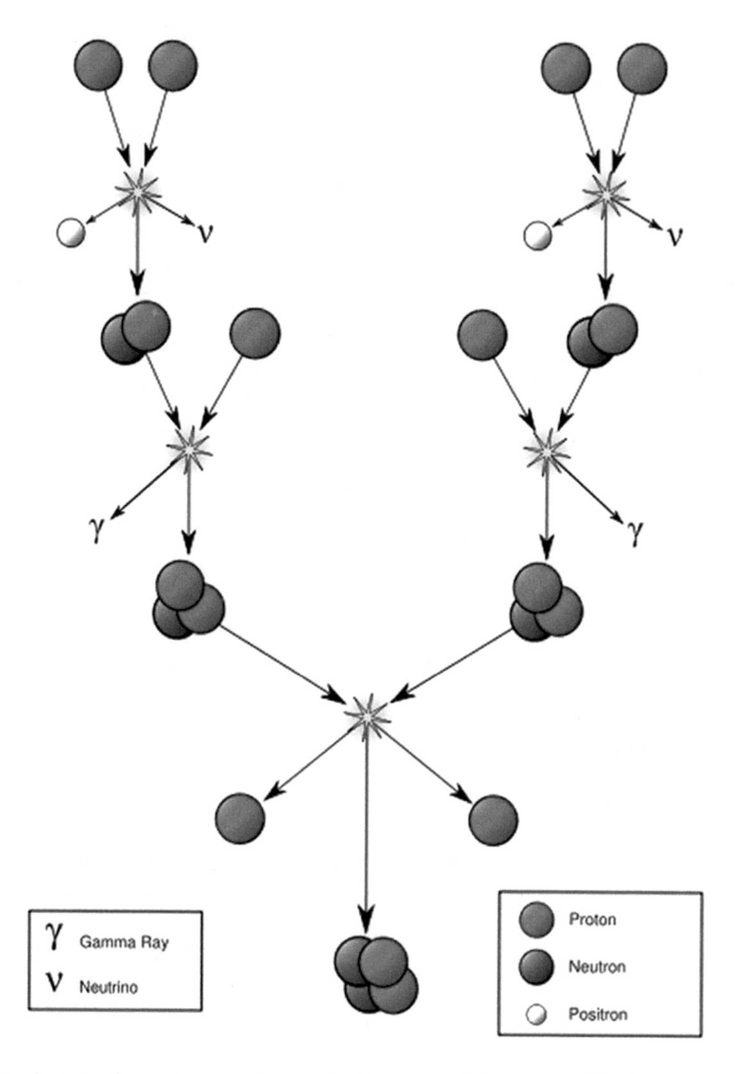

Figure 3.5 In the proton-proton cycle, two mass-1 isotopes of hydrogen undergo a simultaneous fusion and beta decay to produce a positron, a neutrino, and a mass-2 isotope of hydrogen (deuterium) – step 1. The deuterium reacts with another mass-1 isotope of hydrogen to produce helium-3 and a gamma-ray – step 2. Two helium-3 isotopes produced in separate implementations of steps (1) and (2) fuse to form a helium-4 nucleus plus two protons. The net effect is to convert hydrogen to helium, with the energy released going into the particles and gamma-rays produced at each step of the sequence. (Williams College, USA.)

apart under the influence of their own radiation pressure alone. Below this limit, radiation pressure and thermal pressure are in balance with the force of gravity which is trying to cause the star to collapse, maintaining the hydrostatic

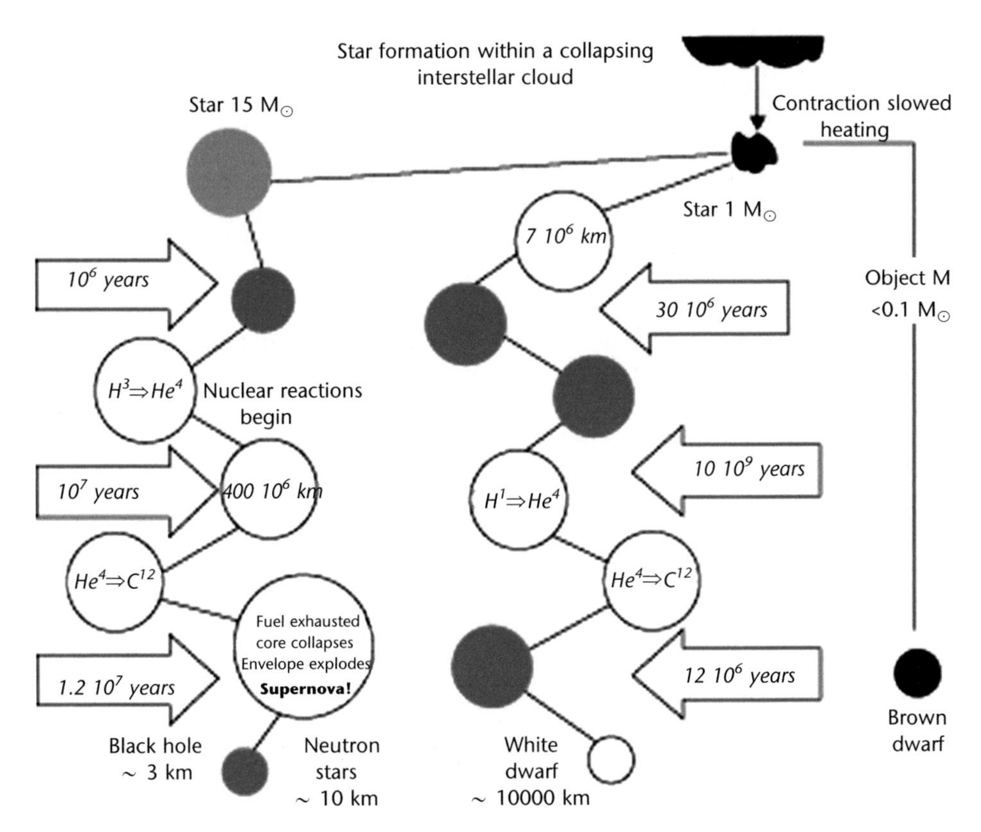

Figure 3.6 Different evolutionary paths of stars as a function of mass, since their formation from a cloud of interstellar matter, within which the Jeans mechanism develops until they end their lives as black holes, white dwarfs or neutron stars.

equilibrium of the layers of the stellar plasma, especially in the star's central regions. However, when for example all the hydrogen in the core is exhausted, the reactions cease, radiation pressure and thermal pressure diminish, and gravitationally-driven collapse resumes.

Lighting the (nuclear) fire

We define a star as a self-gravitating celestial body within which certain conditions are fulfilled which lead to the triggering of nuclear reactions. What are these conditions? To start the process, the distances between the (positively charged) nuclei have to become typically of the order of the dimensions of those nuclei (i.e. 1 fermi = 10^{-15} m), in spite of the force of electrical repulsion. To achieve this, the particles must have very high energy (i.e. temperature); it can be shown that the necessary temperature is so high[6] that no Sun-like star has enough mass to create it.

6. About 100 times the core temperature of the Sun ($\sim 10^7$ K).

To overcome this difficulty and allow the Sun and its siblings to shine, we must appeal to quantum physics, and to a mechanism which sits uncomfortably with traditional physics: the *quantum tunneling effect*. This effect, which is described in more detail in Appendix 4, page 132, 'allows' particles to cross the barrier of electrical potential as if passing through a 'tunnel', allowing interaction between them. In this way, reactions can be triggered at lower temperatures than traditional physics suggests – from 10 to 100 times less, which is just what is needed.

So, hydrogen nuclei can fuse to form helium nuclei in the cores of stars massive enough to fulfill these conditions of density and temperature. The energy liberated by the reaction means that the stars can radiate, yet maintain their equilibrium. This phase, during which stars are in a stable configuration, can last for billions of years. They are then said to be on the *Main Sequence*. The stability of such stars is long-lasting because the effects arising from temperature and pressure constantly balance each other out. The reactions creating the star's energy tend to increase its temperature, which ought to hasten the process; but at the same time, pressure increases with temperature and the dilation of the star due to that internal pressure leads to a decrease in its density, and therefore in its temperature. The rate of nuclear energy production is therefore kept in check, and the system is regulated.

In bodies of lesser mass, the fusion of hydrogen may not even be triggered and, instead of becoming stars, they are 'brown dwarfs' (the existence of which was confirmed by observations of Gliese 229b in 1995). The lowest stellar masses are in fact about one-tenth that of the Sun (Figure 3.7). Below this limit, contraction is hindered by quantum effects (see *Fermi Pressure* in Appendix 2, page 127), and the core temperature never reaches the point where nuclear fusion will begin.

Fundamentally, a star is a *self-gravitating* body, of sufficient mass to allow nuclear reactions to proceed, and its whole life consists of an unceasing battle against the force of gravity.

When a star sees red

Nevertheless, stars do not live for ever, simply because their reserve of combustible material is finite. Instability lies in wait – and the star leaves the Main Sequence.

When the core of the star (about 10 per cent of its total mass) is completely transformed into helium-4, hydrogen fusion reactions cease. The helium atom has two electrons. This means that the force of repulsion between the helium nuclei is greater than is the case with hydrogen. So the helium nuclei cannot fuse at the ambient temperature (several tens of millions of degrees) and the totally inert helium core contracts, as it is incapable of supporting its own weight. The temperature then rises. The gravitational energy thus liberated heats the outer layers (still composed of hydrogen), and fresh nuclear reactions begin as soon as the temperature has reached the critical value.

This 'shell burning' means that the star now has a source of energy peripheral

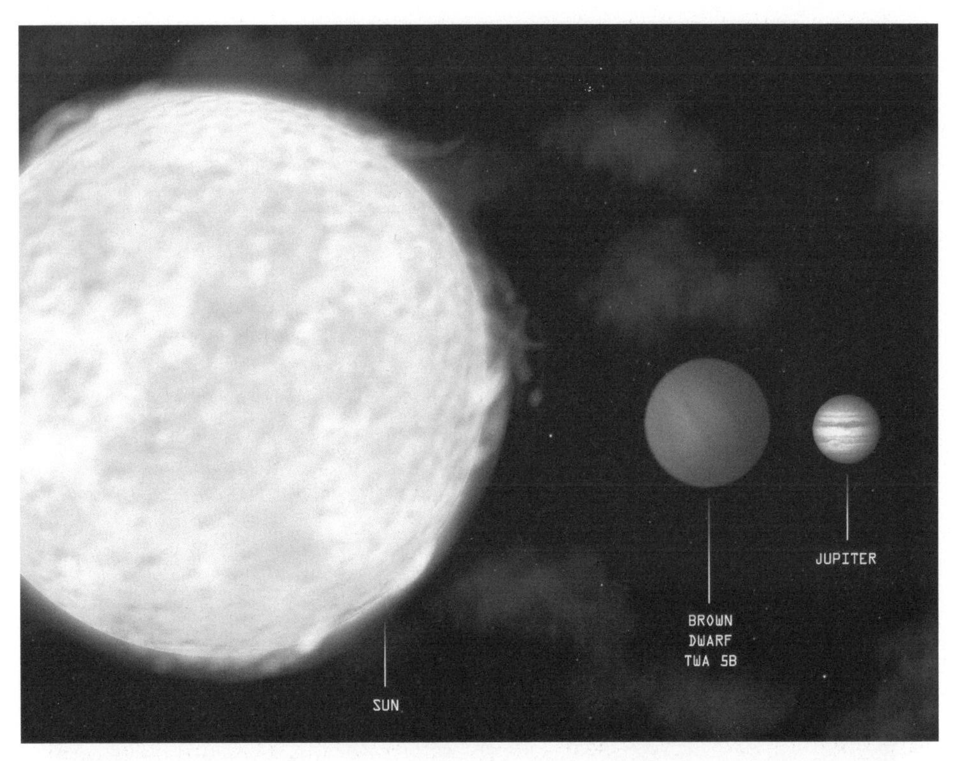

Figure 3.7 The approximate size of a brown dwarf (center) compared to the Sun (left) and Jupiter (right). Although brown dwarfs are similar in size to Jupiter, they are much more dense and produce their own light whereas Jupiter shines with reflected light from the Sun. (NASA/CXC/M.Weiss.)

to the core. This will considerably affect the observed appearance of the star. The energy liberated in this outer layer does not have to support the whole mass of the star, but only that of the exterior envelope, which therefore begins to swell considerably. After a few million years, its diameter may be tens of times greater than it was initially. Its surface temperature will have fallen to about 3 000 degrees, and it will have become a *red giant* (Figure 3.8).

The helium core still continues to contract, its mass being added to by the 'ashes' from the combustion of hydrogen in the outer layer. When the temperature reaches about 100 million degrees, another reaction - the fusion of three helium-4 nuclei into one of carbon-12 - can begin, and stability returns to the core of the star.

During this red giant phase, the appearance of the star has greatly changed: its diameter is much increased, and the star is now red. The greater part of the volume it now occupies is empty space. Throughout 90 per cent of this volume, the density is less than that of the best vacuum we can create on Earth.

Figure 3.8 Expanding light echoes illuminate the dusty surroundings of V838 Monocerotis, a mysterious red supergiant star near the edge of our Galaxy. This image was produced from Hubble Space Telescope data recorded in October 2004. After detecting a sudden outburst from the star in 2002, astronomers followed the flash expanding at the speed of light through pre-existing dust clouds surrounding the variable star. (NASA, ESA, The Hubble Heritage Team (AURA/STScI), and H.E. Bond (STScI).) See also PLATE 12 in the color section.

Growing old gracefully

Eventually, the star runs out of fuel once more, and what happens to it after that depends upon the mass of its central regions. They can resume their collapse, but, if their mass is below what is known as *Chandrasekhar's Limit*[7] (about 1.4 times the mass of the Sun), this collapse will be checked by *degeneracy pressure* (or *quantum Fermi pressure*) created by electrons (see Appendix 2, page 127).

A star having reached this stage is called a *white dwarf*[8] (even though it is essentially now made of carbon). It attains stability, since the degeneracy

---

7. This limiting value is named for the Indian-born astrophysicist Subrahmanyan Chandrasekhar, who formulated it in 1930. He was awarded the Nobel Prize in Physics in 1983.
8. The first white dwarf to be discovered, in 1862, was the companion of Sirius.

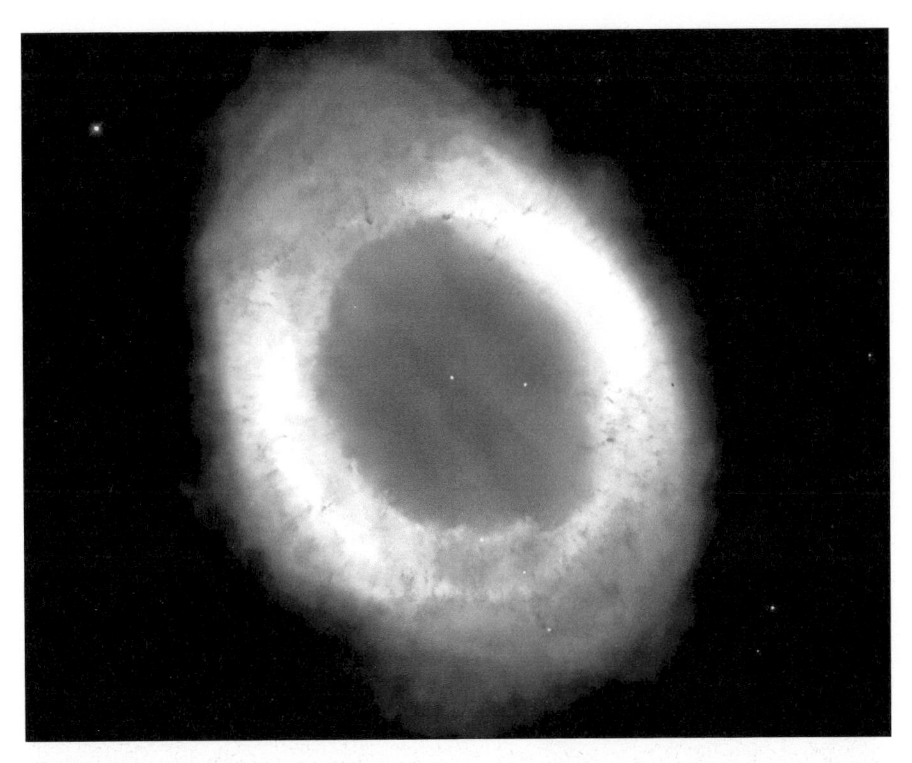

Figure 3.9 A Hubble Space telescope image of the Ring Nebula. The faint speck at its center was once a star of greater mass than our own Sun. Now, near the end of its life, it has ejected its outer layers into space, and the remnant is a tiny white dwarf star, about the size of the Earth. In this image, appearances are deceiving. What looks like an elliptical ring is actually believed to be a barrel-shaped structure surrounding the faint central star, the small white dot in the center. The Ring looks nearly round only because we are looking down the barrel. (The Hubble Heritage Team (AURA/STScI/NASA).) See also PLATE 13 in the color section.

pressure is not dependent on temperature (Figure 3.9). As time passes, the star will gradually cool and grow dimmer, until it becomes a *black dwarf*. These are extremely small and dense objects: their mass is comparable to that of the Sun, but their diameter is more like that of the Earth. Their final years are uneventful.

However, as we shall see below, their sibling stars with cores of masses greater than the *Chandrasekhar Mass* are destined to end their lives in a much more spectacular fashion.

4 Supernova

'If the brightness of the stars were doubled, the universe would be forever dark'

J. W. von Goethe

The explosion of (too) massive stars

An onion with an iron heart

Let us continue to explore the cores of massive stars, stars more than 6 or 7 times as massive as our Sun. In lower-mass stars, the evolution of the central regions is simply arrested when the helium core has been totally transformed into carbon and oxygen; these cores are of insufficient mass to be able to contract any more, with a consequent rise in temperature. So they remain inert, as white dwarfs, held up by degeneracy pressure.

However, the situation is very different in more massive stars. The core can continue to contract, and the internal temperature may reach several hundred million degrees: newly-formed carbon atoms now fuse to create magnesium. The star again achieves equilibrium and core contraction ceases. To an observer, a star in its red-giant phase retains its outward appearance, but in its interior, unseen (and profound) changes are taking place. What we now describe involves only the core of the star, since the outer regions play no part in the star's continuing evolution.

The equilibrium at the heart of the star is, alas, merely transitory. Each time that the combustion of an element ceases and that particular 'fuel' is no longer available, the problem reasserts itself. The star's remedy is always the same: compression of the core in order to increase the temperature, allowing the fusion of elements created in previous stages.

The massive star has now assumed the characteristic structure of a 'nuclear-fusion onion' (as shown in Figure 4.1), with the different reactions occurring in different, concentric layers. The exterior layer consists of hydrogen (H); then come, in order, helium (He), carbon (C), oxygen (O), neon (Ne), sodium (Na), magnesium (Mg), silicon (Si), sulfur (S), and finally, iron (Fe).

The star is, of course, merely putting off the inevitable: the central temperature becomes ever higher, struggling to counteract the ever stronger repulsive forces between nuclei containing more and more electrons. The fusion

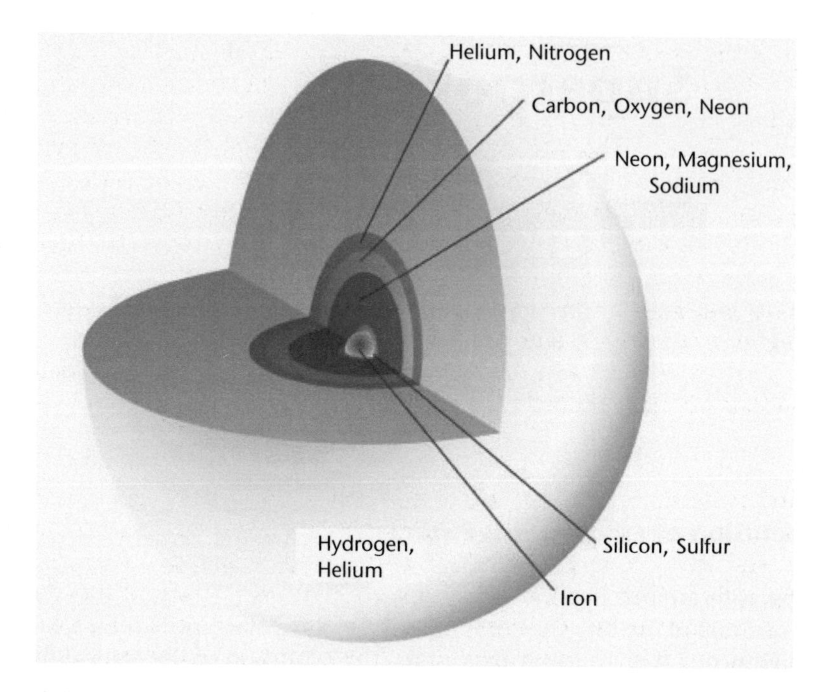

Figure 4.1 The characteristic structure of a massive star, more than 6–7 times the mass of the Sun, showing the 'onion-like' layers representing different phases of contraction and of the ignition of nuclear fusion reactions. The densities and temperatures of the various layers increase with time.

processes creating newer elements evolve more and more rapidly, the rate of these processes being closely linked to temperature.[1]

So, in a star of, for example, 25 solar masses, the reactions proceed at a startling pace:

- combustion of hydrogen to form helium and nitrogen, at a temperature of 30 million degrees takes 8 million years;
- combustion of the helium core, forming carbon, oxygen and neon, at a temperature of 150 million degrees: 500 000 years;
- combustion of carbon to form neon, magnesium and sodium, at a temperature of 800 million degrees: 200 years;
- combustion of oxygen to form silicon and sulfur, at a temperature of 2 billion degrees: a few months;
- and combustion of silicon to form iron, at a temperature of 3.5 billion degrees: just one day.

A headlong flight indeed, and one which can end only in disaster, for the last

1. For example, the rate of fusion in the proton-proton reaction follows a T^4 power law.

element to be created, iron, is the most stable of all natural elements (see Appendix 5, page 134 for more details). At high temperatures, iron will not fuse and produce energy, but will absorb it instead. Iron is the villain of the piece, lurking at the core, ready to engineer the cataclysm which will destroy the star.

This final rapid evolution of the star leaves no time for the enormous outer envelope to react and readjust its structure. Now bloated to a diameter of some 700 million kilometers, the envelope is hardly affected by the death throes of the interior regions. To give some idea of the scale of the structure, if the iron core were 1 centimeter across, the hydrogen-burning shell would be 30 cm across and the diameter of the whole star would be 700 meters.

From now on, we stay with the iron core only, since the rest of the star seems unaware that the rug is about to be pulled out from under its feet.

A heavy heart

When fusion of silicon into iron has ceased through lack of fuel, gravity takes over again, and the core of the star contracts. The innermost region (about 0.8 solar masses) of the core then reaches a temperature of 6 billion degrees. Instead of fusing, as has happened with other elements previously, the iron atoms photodisintegrate[2] under the influence of photons: each iron atom is transmuted into 13 atoms of helium and 4 neutrons are emitted, simultaneously absorbing part of the energy of those photons. This turn of events is, of course, fatal for the star; the fusion which has until now allowed it to resist the force of gravity has now been abruptly switched off. Nothing can now prevent gravity from doing its work ...

In about one-tenth of a second, the innermost zones undergo an enormous collapse, their diameter shrinking from 4 000 km to 60 km; the temperature soars to 50 billion degrees, which is a thousand times greater than that measured at the heart of an atomic bomb explosion. So extreme are the conditions of temperature and pressure that there are now no atoms left in the dying heart of the star: now, only elementary particles (protons, neutrons, electrons, photons and neutrinos) are to be found.

These extreme physical conditions, unparalleled since the beginning of the universe, bring about the *neutronization* of the centre of the dying star, with electrons joining protons to form neutrons. The degeneracy pressure of these neutrons now works efficiently to create a *neutron star*. In this star, matter is so compressed that its density is a million billion (10^{15}) times that of water. Its diameter is between 20 and 60 km, and its surface gravity is about 100 billion (10^{11}) times that of Earth.

---

2. Photodisintegration is a physical process which causes an atomic nucleus to decay into two or more daughter nuclei through the absorption of high energy gamma-ray photons. The process is essentially the reverse of nuclear fusion, where lighter elements at high temperatures combine together forming heavier elements and releasing energy.

Heavy-hearted stars

The core collapse ceases very abruptly after this neutronization phase: the *degenerate neutron gas* now holds its own against gravity. Alas, only a tiny fraction of the star, its former 'heart of iron', is involved in this. The rest of the matter continues to free-fall inwards, eventually to rebound from the extremely dense neutron star. This phenomenon produces a vast shock wave, which propagates through the exterior regions. This shock wave does not pass easily through those peripheral layers of iron which have escaped being neutronized, and which are themselves infalling at a velocity of the order of 70 000 kilometers per second. The encounter is so violent that these iron nuclei, hitherto intact, are in their turn photodisintegrated, taking away a large part of the energy of the shock wave, which will lose all momentum. The first 100 kilometers are the crucial ones in this process. If the shock wave manages to cross what Hans Bethe called 'this minefield', it will have retained enough energy to propagate through the rest of the envelope and cause the star finally to explode. It all depends on the amount of iron encountered in those first 100 kilometers: i.e. on the initial mass of the star. The more massive the star, the larger the iron core will be and the more difficult it will be for the wave to cross the 'minefield'. The upper limit for this to be possible occurs in very large stars of around 20 solar masses, with iron cores of between 1.5 and 2 solar masses.

For a long time, this phenomenon was a major problem for modelers of explosions in massive stars. The outcomes of the models suggested that the explosions would not occur. They did not take account of an important player in the scenario: the *neutrino*. About 10^{58} neutrinos are produced during neutronization, carrying off 99 per cent of the energy emitted during the explosion of the star. Neutrinos do not readily interact with matter, and in the case of stars of initial mass below 20 solar masses, they have no involvement with the phenomenon. However, with more massive stars of much greater internal density, the neutrinos can transfer a tiny fraction (less than 1 per cent) of their vast energy to the outer layers of the star, and this is enough to reanimate the shock wave. The wave will accomplish its mission and escape the danger zone; nothing will be able to stop it (Figure 4.2).

White heat

The encounter between the shock wave and the infalling layers is so violent that *explosive nuclear combustion* takes place, synthesizing the heavy elements later to be found in the interstellar medium, in gas ejected by the explosion. A few hours after the core collapse, the shock wave finally reaches the surface of the red giant, having travelled several hundred million kilometers. The temperature of the outer layers rises from 3 000 degrees to 200 000 degrees. The star now becomes as bright as billions of its fellow stars combined, and as luminous as the whole galaxy in which it was born. For the first time, this phenomenon – involving the ultimate fate of massive stars – becomes visible to the human eye: a supernova is seen.

Hitherto hidden from the astronomers' gaze because the light could not pierce

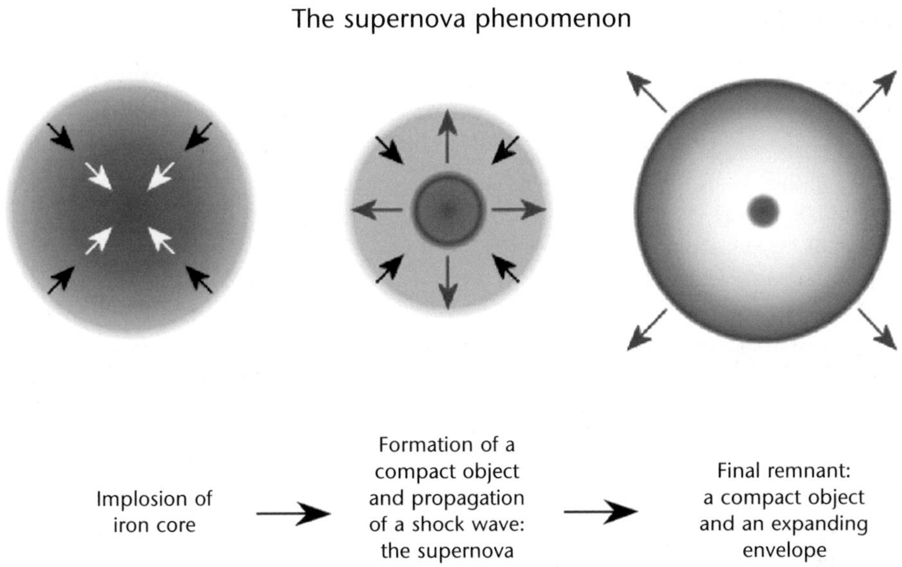

The supernova phenomenon

Implosion of iron core → Formation of a compact object and propagation of a shock wave: the supernova → Final remnant: a compact object and an expanding envelope

Figure 4.2 Principal stages in the formation of a gravitational (core collapse) supernova. The implosion of the core, leading to its neutronization; the propagation of a shock wave in the outer layers; and the explosion of the star (the supernova) leaving a compact remnant (neutron star or black hole) and a rapidly expanding envelope.

the opaque envelope of the star, the drama unfolds: 99 per cent of the energy produced in the explosion disperses into space in the form of a neutrino flux (of about 10^{57} particles). These particles, interacting only very weakly with matter, offer a unique opportunity to gather information on the collapse and evolution of the core of the star. Unfortunately, since neutrinos are barely detectable, large-scale instruments are needed if any are to be 'captured'. To date, only a very few of these precious particles have been detected, principally associated with the recent supernova of 1987 (SN 1987A), relatively close to us in the Large Magellanic Cloud. Most of the shower of neutrinos from this event passed through the Earth without being detected; only a few dozen left any sign of their passage, as luminous rings of *Cherenkov light* within an ad-hoc detector (like the one shown in Figure 4.3).

The amount of kinetic energy liberated from the stellar nucleus is equivalent to that of a billion galaxies like the Milky Way: about 10^{46} joules. Nearly all of this energy is carried by the neutrinos. Only 0.01 per cent of it takes the form of light-energy (see Table 4.1 for the distribution of energy types). Although supernovae offer us a magnificent spectacle, the light that we see represents only a tiny fraction of the total energy involved. The vast majority is lost to the star in a few seconds, carried off by neutrinos.

Now, the implosion of the stellar core does not mean that it disappears altogether. The dense central nucleus survives in the form of a neutron star or a

Table 4.1 Distribution of energy emitted by different types of supernova

	Gravitational (core collapse) supernova	Thermonuclear supernova
Neutrinos	99%	0%
Kinetic energy	1%	99%
Light	0.01%	1%

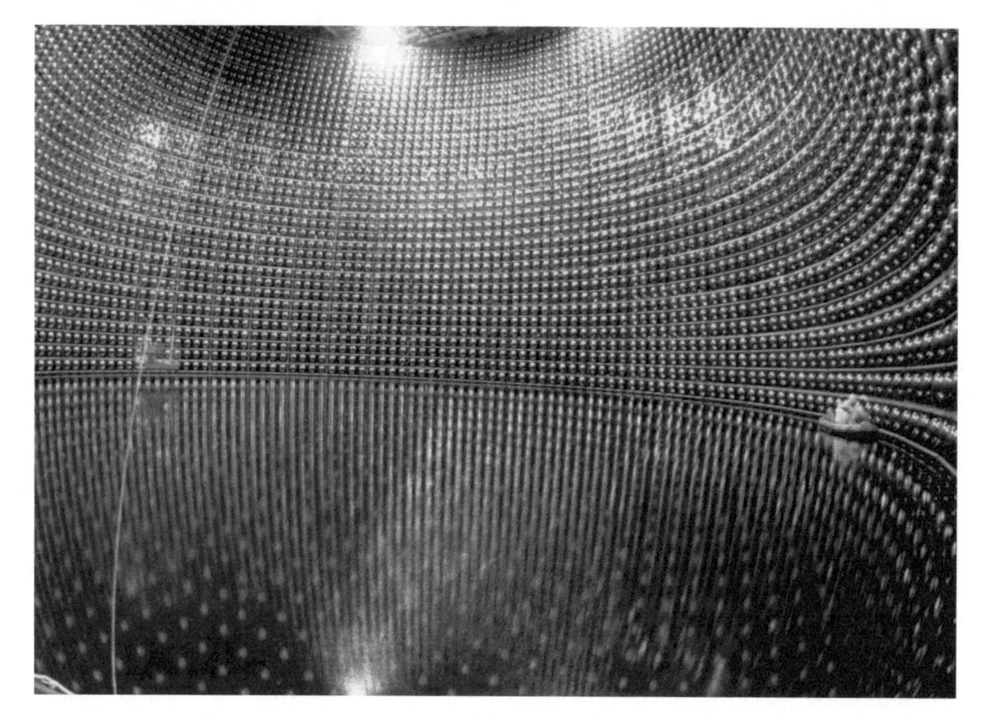

Figure 4.3 The 'Cherenkov Pool' of the Japanese Super-Kamiokande neutrino detector. This is the successor to Kamiokande, which detected neutrinos produced by supernova SN 1987A. Neutrinos are detected by observing the Cherenkov light they create in the pool of pure water. The workers give an idea of the scale of the instrument.

black hole, according to the progenitor star's original mass. As for the matter expelled into space, it will be incorporated into a beautiful supernova remnant, such as the famous Crab Nebula. This object, much admired by observers of the night sky, is shown in Figure 1.2 and in PLATE 2 of the color section.

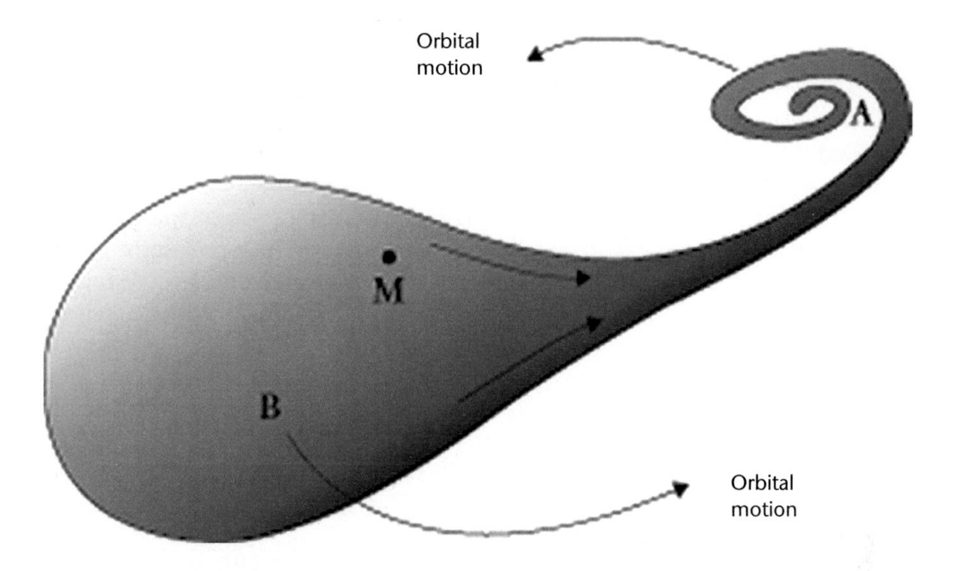

Figure 4.4 The fatal dance of a red giant (B) around its companion, a white dwarf (A). Matter is transferred from one to the other, the whole process culminating in the gigantic celestial cataclysm of a thermonuclear supernova.

Small star wars

It's still mass that counts...

As we have seen in previous chapters, the ultimate fate of stars is linked to their initial core masses. If the mass of the core is less than the Chandrasekhar mass (about 1.4 times the mass of the Sun), fusion ceases after the production of helium, and the star completes its evolution with the creation of a white dwarf, which will cool off over a period of several billion years. If the mass is greater than the Chandrasekhar mass, fusion mechanisms can succeed each other until the star undergoes the ultimate cataclysm of a supernova explosion.

Nevertheless, these smaller stars can also have their moment of glory – if their mass increases, through some external cause, to reach the critical Chandrasekhar mass. Such a configuration is possible in close binary star systems, where two stars are in orbit around each other, and close together (as in Figure 4.4). If the stars are of different masses (which is frequently the case), they evolve at different rates: when the less massive star reaches the red giant stage, the other is already a white dwarf. The bloated and diffuse atmosphere of the red giant is drawn out by the gravity of its companion into a so-called 'Roche lobe'.

Will it or won't it explode?

This mechanism causes the white dwarf to become gradually more massive. It remains stable while the mass is below the critical Chandrasekhar mass, but, as soon as this mass is achieved, the electron gas can no longer support it all. Now, the density of the white dwarf exceeds 2 million kg/m^3, or about a thousand times the density encountered during the evolution of a massive star.

As we have seen in the previous chapter, a white dwarf is an atypical object, composed of degenerate gas. It is important to remember a fundamental property of white dwarfs: pressure within them does not depend upon temperature, but rather upon density. This particular property spells doom for the star: when the temperature is sufficient to allow the fusion of carbon, a large amount of energy is liberated. In a 'normal' star, this surge of energy is regulated by the dilation of the outer envelope; but in a white dwarf, such a mechanism cannot operate because of the degenerate nature of the gas. So a runaway phenomenon begins: the rate of fusion processes increases with temperature, and rising temperature increases the fusion rate... The system literally flies out of control as unregulated thermonuclear combustion builds up, and in a few hundredths of a second, the temperature at the heart of the star jumps from a few hundred million degrees to several billion degrees. Another supernova is unleashed.

But, just like its larger siblings, such a star does not readily explode, and the mechanism triggering the explosion (as well as the details of its evolution) is not well understood. In particular, the combustion of the compact object can be likened to a flame (which is termed a *deflagration*), to an explosion (or *detonation*), or to a 'delayed explosion' involving a phase of deflagration followed by one of detonation. Currently, there is no real consensus as to the nature of the actual mechanism.

In all cases, the combustion of carbon and oxygen in the core of the white dwarf will have converted the atoms into heavy elements, and especially nickel-56, which plays an important role in the stupendous brightness of supernovae. After transforming the central region of the white dwarf, the 'flame' propagates through the rest of the star. As in more massive stars, this propagation does not proceed readily and the flame slows, causing the production of lighter elements such as silicon, calcium, sulfur... After the explosion, no compact object remains: no neutron star, no black hole. All the matter is dispersed into space at velocities exceeding 15 000 kilometers per second.

10^{44} joules of energy are released during the thermonuclear explosion – about 100 times less than in the case of a gravitational or core collapse supernova marking the tragic end of a massive star. However, no neutrinos are produced by this mechanism, and the huge majority of the energy liberated is dissipated as kinetic energy (as shown in Table 4.1).

Family matters

Numbers and letters

The distinction we make between gravitational or core collapse supernovae, linked with massive stars, and thermonuclear supernovae, linked with white dwarfs, is a comparatively recent one. Although these terms are well suited to the observed mechanisms, astronomers have kept to a more traditional nomenclature which has evolved with our study of supernovae.

In 1940, Rudolph Minkowski discovered a supernova which was very different from the twenty or so already known. For the first time, the light curve (representing the evolution of luminosity as a function of time) was noticeably different from previous examples, and hydrogen lines were very obvious in the spectrum.[3] This last observation was no surprise, since hydrogen is the most abundant element in the universe and the main component of all stars. Astronomers had been astonished when they found no trace of it in the spectra of supernovae.

Minkowski's observation showed that supernovae were not all part of a homogenous family. It became necessary to sort them into classes reflecting this new demographic:

SN Type I: no hydrogen lines.
SN Type II: hydrogen lines present.

Urged on notably by Fritz Zwicky, research projects led to the discovery of various other types of supernova, and inevitably, as differences became apparent, so sub-classes arose (examples of their spectra can be found in Figures 4.5 and 4.6):

SN Type Ia: no hydrogen lines, silicon present.
SN Type Ib: no hydrogen lines, strong helium presence.
SN Type Ic: no hydrogen lines, no helium, weak silicon lines present.

Another important observation helps us to understand the origin of supernovae. No supernova of Type II, Ib or Ic has ever been observed in an elliptical or lenticular galaxy, but they are common in spiral galaxies. Conversely, Type Ia supernovae explode in both spiral and elliptical galaxies. This is a very important fact, since it leads us to conclude that:

- supernovae of Types II, Ib and Ic are associated with galaxies whose stellar populations are relatively young;
- supernovae of Type Ia are associated with relatively aged stellar populations.

These observations, together with advances in theories of stellar evolution, now open the way to a coherent classification. Type Ia supernovae result from the thermonuclear explosions of white dwarfs. This mechanism explains the absence of hydrogen lines in the spectrum since the white dwarf is essentially composed of carbon and oxygen. White dwarfs evolve from low-mass stars such

3. The distribution of energy as a function of the frequency of the light, or more generally, of the electromagnetic wave detected.

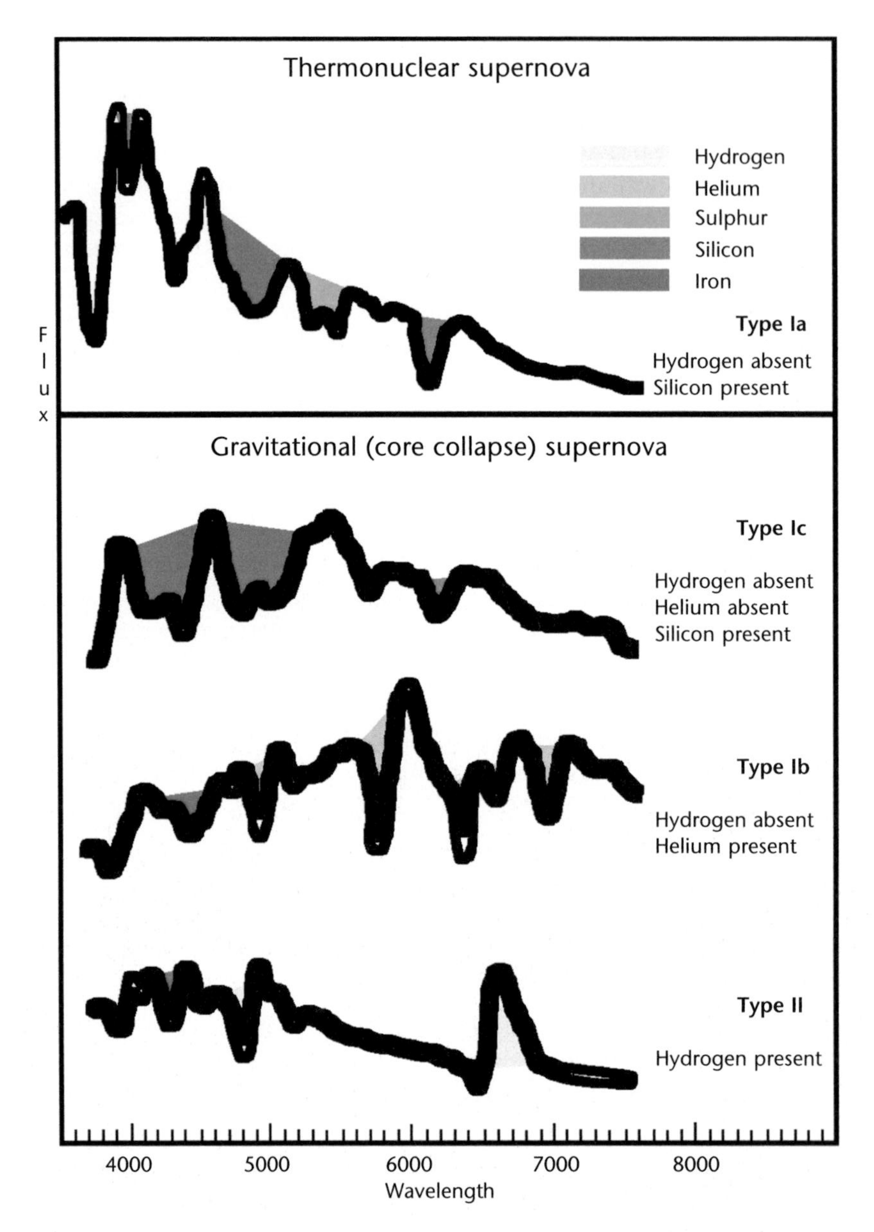

Figure 4.5 Principal differences in the spectra of supernovae. Although they differ in detail and vary over time within the same class, these spectra can be classified into two main fundamental categories according to whether or not they exhibit hydrogen lines. Unfortunately, this classification does not correspond completely to the reality of the phenomenon: thermonuclear supernovae are associated only with Type Ia, whereas gravitational (core collapse) supernovae are associated with Types Ib, Ic or II.

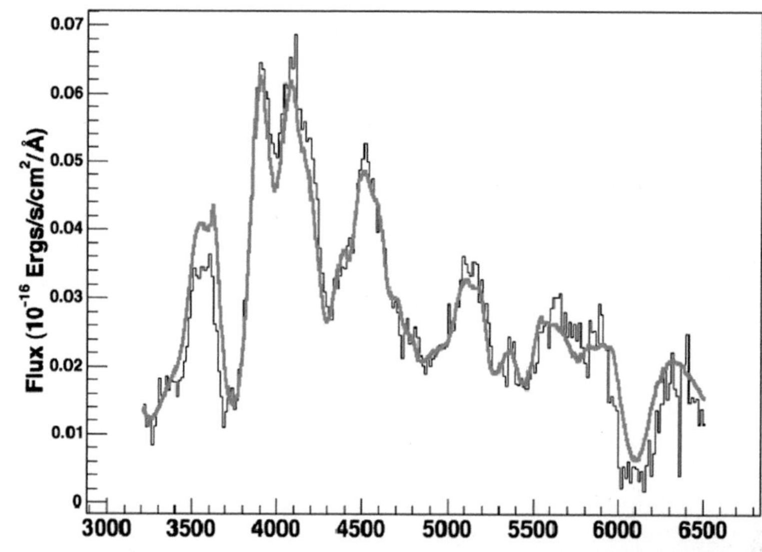

Figure 4.6 (a) Example of a Type Ia thermonuclear supernova spectrum detected and monitored by the Supernova Legacy Survey Program. With a redshift of only 0.3, this supernova is 'relatively' close to us. It exploded only 3.4 billion years ago! **(b)** The observations were carried out by the Very Large Telescope (VLT) of the European Southern Observatory (ESO) on the summit of Cerro Paranal in Chile. See also PLATE 14 in the color section.

as our Sun, and are therefore relatively abundant in all types of galaxies. It is therefore natural to find Type Ia supernovae in all types of galaxies.

Supernovae of Types II, Ib and Ic result from the gravitational collapse of massive stars. The absence of hydrogen or helium in the spectrum is due to radiation pressure, which can be particularly assertive in high-mass stars (as detailed in Appendix 3, page 130). This pressure can drive off the hydrogen envelope, leaving the helium (Type Ib), or drive off both envelopes (Type Ic). Since the lifetime of a high-mass star is relatively short (just a few million years), we expect to find these supernovae only in galaxies where star formation is still going on.

Light curves

The brightness of all types of supernova increases for about a fortnight or so after the explosion, then gradually falls off as the months pass. Nevertheless, the evolution of the brightness differs depending on whether the supernovae are gravitational or thermonuclear (as shown in Figures 4.7 and 4.8).

With gravitational or core collapse supernovae, the emission of light is due to the heating of the outer layers by the shock wave generated by the collapse of the iron core. First, the layers dilate and their surface area increases rapidly. The

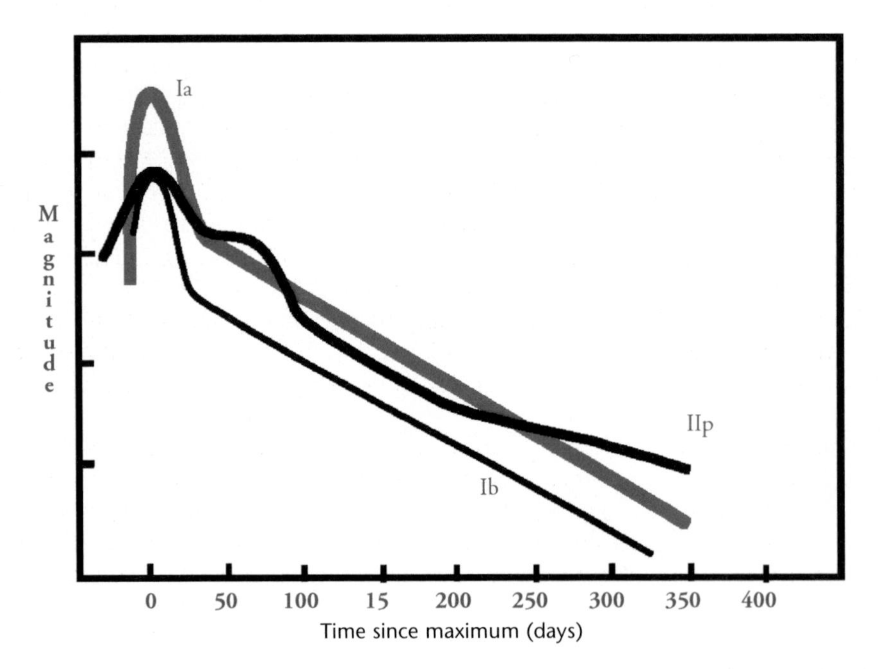

Figure 4.7 The expected evolution of luminosity over time for different types of supernova. The luminosity increases rapidly, reaching its maximum in 3-4 weeks, thereafter decreasing over several months. The shape of the light curve is different according to the type of supernova.

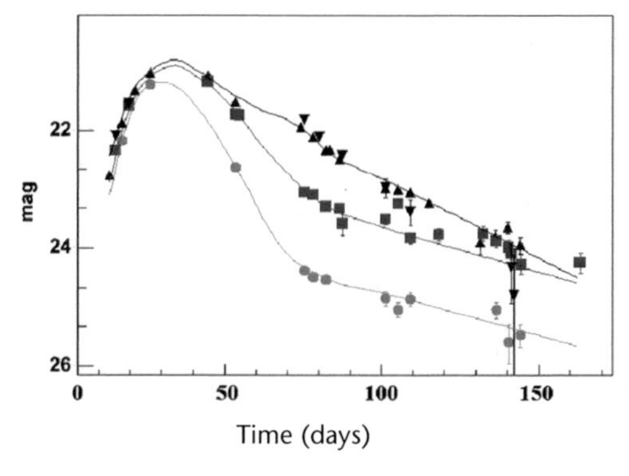

Figure 4.8 The light curve of a Type Ia thermonuclear supernova, detected and monitored by the Supernova Legacy Survey Program. The observations were carried out by the Canada-France-Hawaii Telescope (CFHT) on the summit of Mauna Kea in Hawaii. The curve was obtained with the MegaCam imager, using several filters, from blue (lower curve) to red (upper curve).

brightness of the supernova, which is proportional to this surface, begins to increase progressively. Three or four weeks later, the expansion of the envelope, which has compensated for the cooling effect, can no longer maintain the brightness of the supernova, and it will have achieved its maximum brightness.

When the temperature of the outer layers reaches 5 000 or 6 000 degrees, an important phenomenon supervenes: electrons and protons, which have until now been separate, can recombine to form atoms. The elements of the (cooler) outer part of the upper layers are once more atomic in nature, but those further down in the interior are still hot and ionized. Since atoms allow the transmission of light, the outer layers become transparent while the internal layers remain completely opaque. The transitional zone between these two domains, each with its own particular properties, is known as the *photosphere*. This imaginary 'surface', the temperature of which is well known, remains at a constant distance from the supernova for a while. The brightness of the supernova shows little change for a few months, creating the characteristic plateau in the light curves of Type II supernovae.

An unexpected consequence of this phenomenon is that we 'see' ever deeper layers of the supernova paraded before our eyes. The layers, driven by the shock wave, become directly observable as they reach the level of the photosphere. By studying the temporal evolution of the spectra, we can examine the different layers in detail. This is the only opportunity to perform a stellar 'tomography', looking deep into these stars and observing their famous onion-like structure.

Enter radioactivity

Just as in the case of the gravitational or core collapse supernova, the brightness of a thermonuclear supernova increases initially due to the very rapid expansion of the small envelope around it. This envelope is much smaller than that of a massive star, and we might unsuspectingly imagine that the brightness of a thermonuclear supernova should be well below that of a supernova of gravitational origin.

But thermonuclear supernovae have a 'secret weapon': radioactivity. As the white dwarf burns, a considerable fraction of its mass is transformed into very unstable radioactive nuclei of nickel-56, which disintegrates into cobalt-56 while emitting gamma-rays. The gamma-rays will interact with free electrons and transfer some of their energy to them. This causes the excitation of intermediate nuclei such as oxygen, silicon, sulfur and calcium. The de-excitation of these atoms adds to the luminous output of the supernova, and its total brightness is increased. Maximum brightness will occur about two or three weeks after the explosion. At its maximum brightness, a thermonuclear supernova is 3 to 6 times brighter than its gravitational counterpart. Thereafter, brightness decreases according to a regular rhythm, and the amount of nickel-56 becomes progressively smaller. About two months after maximum, almost none of the nickel remains; it is now the turn of cobalt-56 to take up the baton. This element will similarly disintegrate, to produce iron-56 and gamma-rays. Since cobalt-56 has a lifetime ten times longer than that of nickel-56, it reinvigorates the atoms of the supernova and slows the rate of decrease in its brightness.

Fireworks in the sky

Although the mechanisms powering both thermonuclear and gravitational supernovae are very different, they occur with approximately the same frequency.[4] The only appreciable difference is that gravitational or core collapse supernovae, developing from massive stars, appear only in 'young' spiral galaxies.

Within our own Galaxy, it is estimated that one to three supernovae explode every century. Such a luminous phenomenon is unlikely to pass unnoticed, and the history of humanity ought to be littered with observations. However, the disc of our Galaxy is relatively dusty, and the light from most supernovae is very readily absorbed. This is the reason why the six observed historical supernovae (Table 1.1) are situated in our galactic neighborhood, which might be said to encompass only about 20 per cent of the volume of the Galaxy.

The frequency mentioned may seem fairly low on the scale of a human lifetime, even if veritable 'supernova factories' do exist in the cosmos (such as the galaxy in Figure 4.9). However, if we take into account the fact that our universe comprises billions of galaxies, we conclude that about five supernovae explode every second in that universe. Fireworks indeed.

---

4. The supernova rate measurement is a difficult exercise. However, we can estimate that the ratio is roughly two gravitational (core collapse) supernovae to each thermonuclear one.

Figure 4.9 The Fireworks Galaxy NGC 6946. This is a relatively nearby, face-on spiral galaxy to our Milky Way, located just 10 million light-years away towards the constellation of Cepheus. Looking from the bright core outward along the loose, fragmented spiral arms, there is a change from the light of old stars in the galaxy's center to young star clusters and star forming regions. NGC 6946 is rich in gas and dust, with a high star birth and death rate. In the past century, at least eight supernovae were discovered in this galaxy. Their positions are marked here. See also PLATE 15 in the color section.

Where is the cosmology in all this?

Supernovae are very fascinating objects. They represent the most violent phenomena since the very formation of our universe. But the most extraordinary thing is that these objects, as will be revealed in some detail in the following chapters, have become the favored 'skymarks' for surveying and studying the cosmos. This is a fairly recent development and arises from the fact that supernovae are intrinsically brilliant objects, which means that we can observe them at great distances.

Moreover, since the 1990s, we have realized that Type Ia (thermonuclear) supernovae can be used as 'standard candles'. This property, although still not completely understood, is mostly due to the fact that this phenomenon occurs when a white dwarf reaches the Chandrasekhar mass.

However, before embarking upon the cosmological applications of supernovae, we must turn our attention to an extreme phenomenon of stellar evolution: gamma-ray bursts.

5 Supreme stars: gamma-ray bursts

'And in their turn, the stars mark the demise of the day, like funeral torches.'
Jean Racine

An amazing menagerie

From the time of their discovery in the early 1970s until their still only partial elucidation in 1997, the origin of gamma-ray bursts (GRBs) was shrouded in great mystery. For a long time, this phenomenon had been a real puzzle, and many and varied were the rival theories put forward to explain it. Some of these theories proposed mechanisms within our own Solar System; others ranged through our galaxy; and still others looked beyond it. It was all a question of size, given the observed energy levels: if indeed this were a phenomenon at cosmological distances, it would be one of the most violent known in our universe, and could even outstrip the supernovae in the discovery stakes.

The veil was partially lifted by the American Compton Gamma-Ray Observatory (CGRO) satellite, launched in 1991 (Figure 5.1). One of its instruments, BATSE (Burst and Transient Source Experiment), consisting of eight gamma-ray detectors, observed more than 2 700 gamma-ray bursts (see Figure 1.10), providing detailed information about these objects: their light curves (flux as a function of time), spectra (distribution of energy as a function of frequency), and positions to within a few degrees.

Short, long, but all different

A first and very interesting result arising from the study of the light curves (Figure 5.2) was that they exhibited a great variety in behavior. On the one hand, the duration of the emissions varied, from a fraction of a second to several tens of seconds, and sometimes several minutes; on the other hand, variations in intensity on timescales as short as a few milliseconds were observed.

These variations were the first indication that the object within which this burst of gamma-rays originates is extremely compact. However, the great diversity hindered the classification of these bursts on the basis of their timescales alone.

Eventually, the time interval during which 90 per cent of the photons are detected, for the totality of known bursts, was considered, and now the

Figure 5.1 The Compton Gamma Ray Observatory (CGRO) satellite with solar panels unfurled is pictured here just prior to its release into orbit after deployment from the payload bay of the Space Shuttle. Four of the eight detectors of its BATSE (Burst And Transient Source Experiment) instrument are seen at the four corners of the satellite. (NASA)

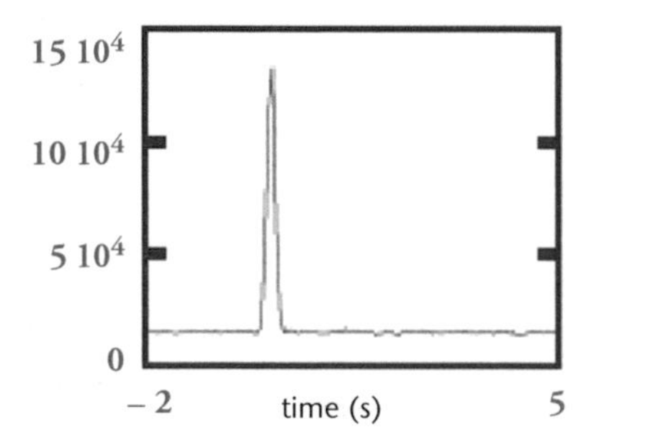

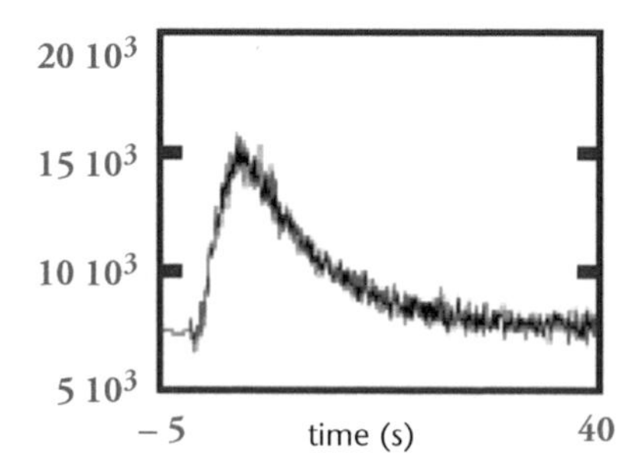

Figure 5.2 Example of light curves obtained by BATSE, on board CGRO. The horizontal axis represents time in 'seconds since trigger', and the vertical axis is proportional to the number of photons received. Actually, in gamma-ray astronomy, photons are very energetic but very rare: the instruments detect them one by one.

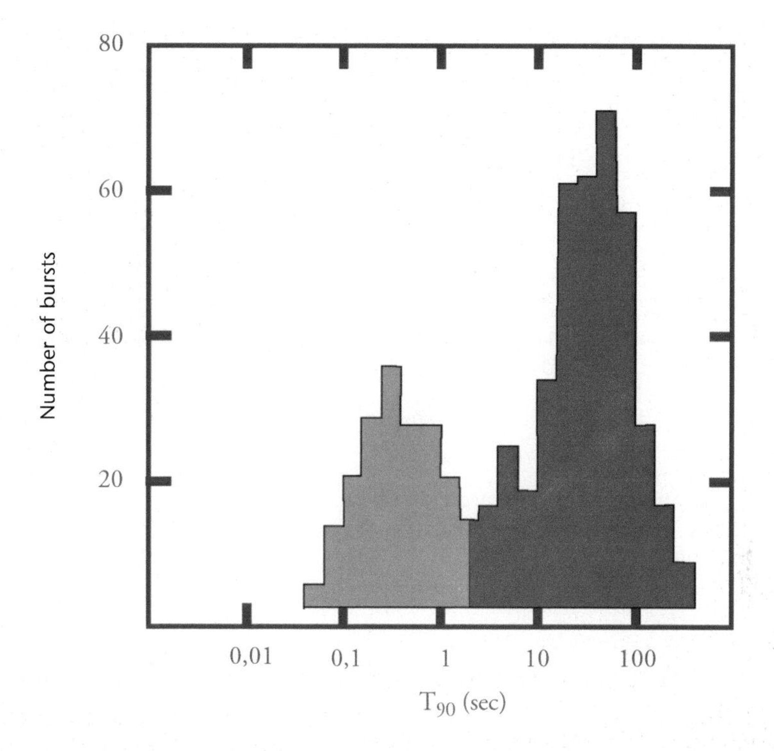

Figure 5.3 Distribution of time intervals during which 90 per cent of the photons were detected by BATSE. The existence of two distinct families is very obvious: short bursts (light grey) and long bursts (dark grey).

distribution of the durations of the events became clearly separated into two distinct families (as shown in Figure 5.3).

This first, particularly intriguing, finding led to the classification of the bursts into two categories.

- Long bursts, with emissions lasting more than two seconds. These represent about three-quarters of GRB events.
- Short bursts, with emissions lasting less than two seconds.

The first findings based on the BATSE observations caused deep consternation in the scientific community. It would not be easy to find a single mechanism to explain such a bimodal distribution. Might two different mechanisms be a more likely explanation? Before going into the question of the origin of gamma-ray bursts, we must delve further into their properties, in the search for other clues left by the culprit(s).

Exuberant energy

The energy distribution in the signal from gamma-ray bursts holds important

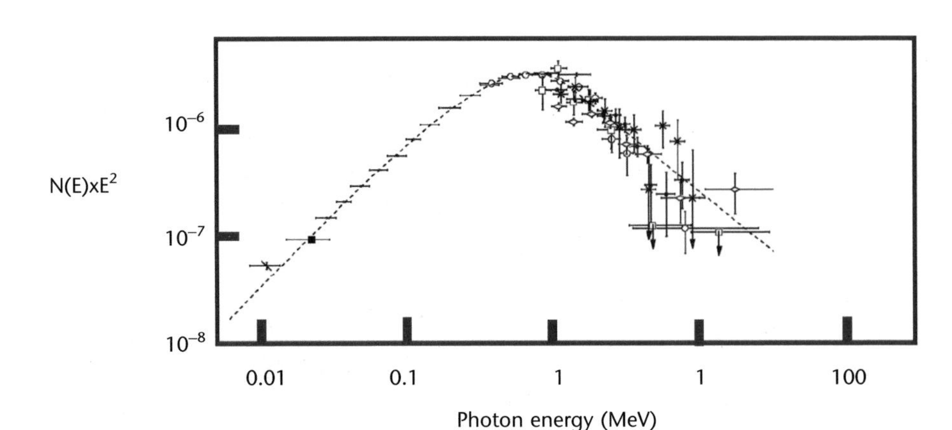

Figure 5.4 Example of the spectrum of gamma-ray burst GRB 990123, in a diagram with $N(E) \times E^2$ as a function of the energy between E and $E + dE$. From this representation we clearly see that there is a maximum (E_{peak}) located, in this example, at about 720 keV. In reality, this is an extremely energetic burst, with its energy typically peaking between 150 and 230 keV.

information on the nature of the phenomena involved. The distribution suggests that they emit nearly all their energy in the gamma-ray domain.[1]

This property is apparent when we assess the number of photons $N(E)$ of a gamma-ray burst as a function of energy, in a diagram investigating the quantity $N(E) \times E^2$. We see from this representation that the spectra have their maxima for each burst, labelled E_{peak}, indicating the spectral band where the *peak energy* of the object is located. The profile of these spectra can be readily replicated by a function which can be broken down into two parts with reference to the peak energy (an example of this is shown in Figure 5.4).

The form of these spectra strongly hints that the mechanisms responsible are in reality 'non-thermal', i.e. independent of temperature, contrary to what is observed within stars or supernovae. They rather indicate some process involving *synchrotron radiation.*

The informed reader will have noticed that we have not made any distinction between long and short bursts. In fact, their energy spectra are similar, even though the value of their energy peak E_{peak} is lower in the case of the long bursts. The value for long bursts is about 150 keV, while the value for short bursts is about 230 keV. Short bursts are therefore slightly more energetic than long ones.

---

1. Assume in this context that in any reference to optical or radio emissions (undetected during the BATSE observations), those emissions remain marginal in comparison with the gamma-rays.

Bursts from all directions

A final, and particularly striking, result from BATSE observations is the chart of positions of gamma-ray bursts in the sky. The distribution of these across the whole sky shows a clear *isotropy*. No favored direction has been detected, nor any tendency to clustering on either a small or a large scale. We use certain criteria in order to verify quantitatively that there is no correlation between the positions of these bursts across the celestial sphere, and that of our Galaxy. The first criterion involves the angle between the burst and the Galactic Centre: do the bursts tend to concentrate towards that centre? The second involves the Galactic latitude of the direction in which the bursts originate: do they tend to lie along the plane of the Galaxy? If we apply these criteria to the BATSE observations, the result is unambiguous. The directions from which the bursts originate are in statistical agreement with the hypothesis of a perfectly isotropic distribution. This means that we can eliminate any scenarios suggesting that these objects are within our Galaxy, unless they are very local events (a hypothesis only much later disproved, as we shall see).

These results, at first disconcerting, nevertheless constitute a remarkable advance in the characterization of the phenomena involved in gamma-ray bursts. As for actually understanding them, we should not have too long to wait.

A sad fate

From the edge of the cosmos

One of the key questions about this phenomenon involves distance. If gamma-ray bursts are indeed cosmological in origin, they represent nothing more or less than one of the most violent phenomena in the universe, surpassing supernovae in this strange cosmic menagerie. It was not until 1997 that the uncertainty was resolved, thanks to a remarkable technical feat involving the Italian-Dutch satellite BeppoSAX. On board, a telescope equipped with X-ray detectors was repositioned to point in the direction of a gamma-ray burst, just a few hours after a discovery alert had been issued.

This achievement, a real 'first' for BeppoSAX, associated a long gamma-ray burst, GRB 970228,[2] with an object emitting in another domain of the electromagnetic spectrum. This observation showed that GRBs emitted signals in a very wide spectral domain, and that their location might therefore be pinpointed to within a few minutes of arc.

It was therefore possible to direct telescopes operating in visible light towards this object, and a light source, fading over a timescale of a few days, was indeed

---

2. The nomenclature used for GRBs is based on their discovery dates. So, for example, GRB 970228 was detected on 1997 February 28 (year-month-day). This system is very like that used for supernovae. Perhaps astronomers sometimes lack imagination...

observed. This phenomenon, known as an *afterglow*, radiated at frequencies from visible light to X-rays.

This technical feat opened the way to accurate measurement of the distance of these newcomers to the cosmic menagerie. A few months went by, and then another afterglow spectrum, that of gamma-ray burst GRB 970508, was discovered. On this spectrum, several sets of absorption lines were clearly visible. These lines were due to the presence of matter between the observer and the afterglow. The burst was, necessarily, at least as far away as the absorbing material, and was therefore undoubtedly at a cosmological distance.

These new findings gave closure to a debate which had lasted nearly 30 years: long gamma-ray bursts were well and truly of cosmological origin. Another question now arose: the intrinsic brightness of these phenomena was absolutely enormous. They dethroned supernovae, quasars and other monsters of Nature, to become the most energetic events since the formation of the universe itself. To date, there have been many observations leading to measurements of the distances of GRBs.

Up to the time of the final reading of the English language edition of this book, the record holder was GRB 050904 (detected on 4 September 2005). Its spectral redshift tells us that it occurred when our universe was less than a billion years old, at a time when our own Galaxy had not yet formed. This is a truly remarkable observation, since it reveals to us one of the most distant astronomical objects ever seen.[3] Its spectrum is shown in Figure 5.5, and its visible-light counterpart in Figure 5.6.[4] For the purposes of comparison, the most distant supernova yet detected has a redshift of 1.7, separating it some 3 billion years in time from GRB 050904.

Gamma-ray bursts and supernovae: an unexpected kinship

In 2003, there was another development, involving the observation of a new, long gamma-ray burst, GRB 030329. This event is currently one of the nearest yet observed. Its redshift of only 0.168 suggests that it occurred only 2 billion years ago. About ten days after this discovery, and to the great surprise of astrophysicists, a supernova appeared in exactly the same place as the gamma-ray source, showing for the very first time that there was a close link between gamma-ray bursts and supernovae.

Since then, other GRB-linked supernovae have been observed. In every case,

---

3. It is probable that the Swift satellite, designed to study GRBs, has already observed events with redshifts greater than 10, placing them at a time when the universe was less than 500 million years old. Unfortunately, it has not yet been possible to confirm this directly.
4. The record held by GRB 050904 has now been beaten. The new record holder is GRB 080913, detected on 13 September 2008, during the final reading of the English language edition off this book. It is about 70 million years younger than GRB 050904; a small step for mankind – a giant leap for science.

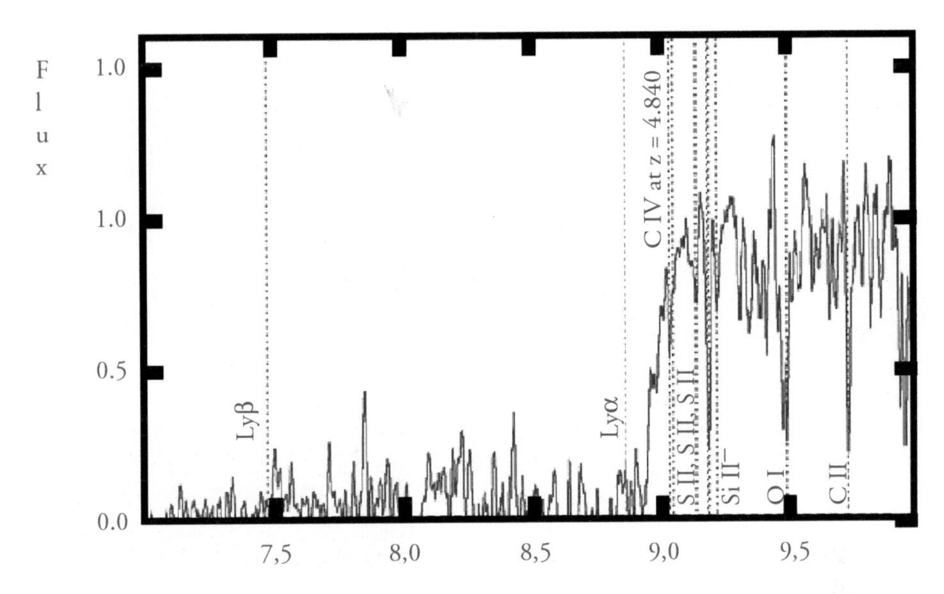

Figure 5.5 Spectrum of one of the most distant gamma-ray bursts ever observed: GRB 050904. Its redshift is at least 6.29, placing it among the most distant objects. This spectrum was obtained by the Japanese Subaru telescope on Mauna Kea, Hawaii.

these are 'gravitational' supernovae, resulting from the core collapse of a massive star which has already exhausted its hydrogen (Type Ib) and helium (Type Ic).

So, it seems that long gamma-ray bursts are linked with the deaths of massive stars.

This astonishing connection is corroborated by studies of the properties of the host galaxies of these objects. It does seem that these galaxies are undergoing a particularly active phase of star formation, with between 1 and 10 stars of the same mass as our Sun being born every year. Such an environment encourages the formation of massive stars, which may well, after only a few million years, become gravitational (core collapse) supernovae.

Short GRBs

It took more time to explain the origin of short gamma-ray bursts, since no afterglow had yet been detected, in either the X-ray or the visible domains. It had therefore not been possible to fix their positions accurately. Only seven years after the first long GRB distance estimation, in 2005, was any light thrown upon their origin when, as with their longer-lasting counterparts, the anticipated afterglow was finally observed – first in the X-ray domain, and then in the visible. These observations led to the identification of the host galaxies, and unequivocal distance measurements could be made.

So, short gamma-ray bursts also take place at cosmological distances, though it was established early on that they are mostly much nearer than their longer-

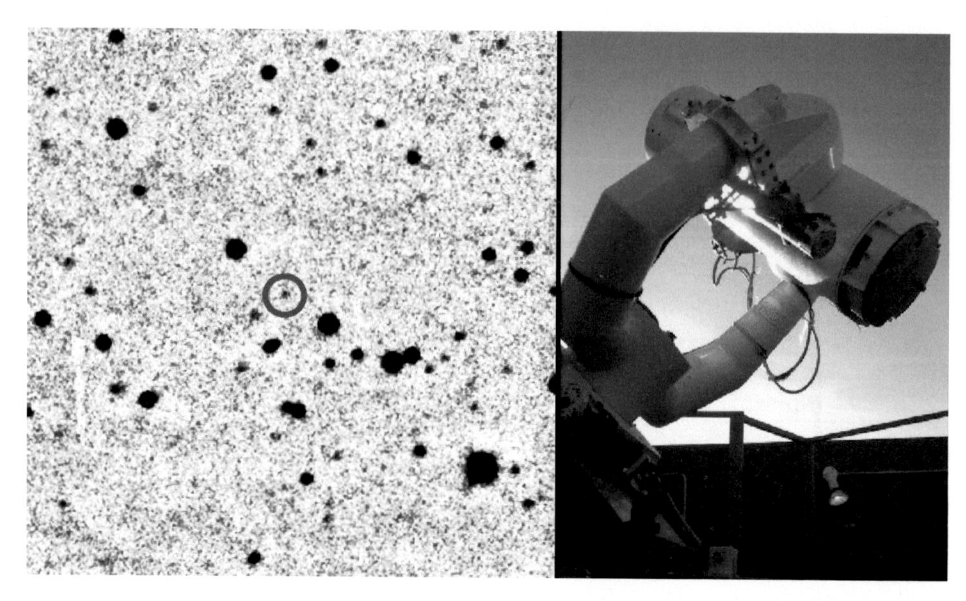

Figure 5.6 Observing the distant gamma-ray burst GRB 050904 in the optical domain, with the TAROT robotic telescope in France. This represents a real observational feat, the telescope being only 25 cm in diameter! The extreme luminosity of the phenomenon made observation possible for about 10 minutes, before the intensity of the source decreased and telescopes of greater aperture were required.

lasting siblings.[5] At present, the average redshift for short bursts is around 0.3, with the most distant examples occasionally reaching 0.7, while the most distant long burst shows a redshift of 6.7. The conclusion is that the short bursts we see nowadays are relatively recent phenomena in the history of our universe, having occurred some time during the last 7 billion years.

Having established a link between long bursts and supernovae, it was natural that similar associations should be investigated in the case of short bursts. Many studies have been carried out, but none of them has yet borne fruit. It has not been possible to connect short bursts with the deaths of massive stars. This conclusion has been corroborated by studies of the properties of the host galaxies of short GRBs. They are all occurring within galaxies which exhibit almost no star formation, and in which massive stars have long since disappeared, billions of years ago. Short bursts cannot therefore be linked with gravitational supernovae, which are the products of very massive stars at the ends of their very short lives.

5. This is of course relative, since we are dealing here with matters of minutes rather than the billions of years representing the lifetimes of ordinary stars.

Each to his own fate

Historically, gamma-ray bursts have been identified as short-lived and intense flashes of gamma-rays. Various observations have, in the course of time, led us to think that the celestial body within which they originate, the *progenitor*, is different in nature according to whether the burst is short or long, though the phenomenon evolves consistently thereafter for both types.

The task therefore becomes the identification of the exact nature of this progenitor, and its formation. One element of the answer has been supplied by studies of the energy emitted during this phenomenon, energy exceeding 10^{43} joules: the equivalent of the energy dispersed by a star such as our Sun over a billion years. The cradle of all this activity is moreover extraordinarily compact, as shown by the very great variability of the flux of the radiation over timescales as short as a millisecond. The only body compact enough to be able to convert into radiation the enormous quantity of matter required is ... a black hole.

To form such a body, short and long bursts would have to follow very distinct pathways.

A beneficial pairing

The *danse macabre* of short bursts

The most widely favored hypothesis for the formation of short bursts is based on the fusion or 'coalescence' of binary systems comprising two compact objects: white dwarfs, neutron stars or black holes. Such systems have already been encountered in the previous chapter, in the explanation of the genesis of thermonuclear supernovae. It is a fact that more than half of all stars observed are not alone, but belong to double-star systems. The life of the stellar couple, be it calm or stormy, is governed by only two parameters: their masses, and the distance between them.

First of all, it is mass which controls the evolution of each star. It determines which fuel source they will be able to burn; the lengths of their lives; and finally, the manner of their deaths. Binary systems which initially comprise two massive stars will evolve, over timescales of between a few million to several billion years, into pairs of compact objects which may be white dwarfs, neutron stars or black holes.

The second parameter is the distance between the two compact objects. In the case of a close binary system, the orbital characteristics of the pair will eventually be modified due to the emission of gravitational waves, and the distance between them will decrease over timescales of between a few tens of millions of years to a billion years.

Spiralling ever inwards, the two objects will finally coalesce, and the mass of this new system is sufficient to create a single black hole. A vast amount of energy is thereby liberated in a short period, with a brief but intense flood of gamma-ray emissions: the gamma-ray burst. The mechanism which now unfolds is very similar to that already encountered in the case of long bursts. It will be described below.

This scenario, with two compact objects coalescing, has merit in that it broadly explains the properties of short bursts. In particular, since the formation and coalescence of a compact binary system takes a long time (several billion years), it is natural that we see this process late on in the history of the universe, in evolved galaxies where star formation has ceased.

It is a fascinating thought that, since this mechanism will inevitably give rise to gravitational waves, their possible detection by existing instruments such as VIRGO in Italy and LIGO in the USA, or by future projects such as LISA in space (Figure 5.7), would provide striking proof of the reality of the proposed scenario. This is an exciting prospect for twenty-first century astronomy.

A star too big: long bursts

The observation that long bursts are associated with gravitational (core collapse) supernovae leads us naturally to link this phenomenon with the final stages of stellar evolution in the most massive stars. Such stars, of masses exceeding 20-30 times that of our Sun, end their lives as black holes. The black hole grows rapidly, swallowing up the matter in its vicinity while simultaneously emitting, in some cases, two relatively narrow jets (with an opening angle of a few degrees) in opposite directions.

The jets encounter some difficulty in escaping from the extremely massive envelope of the supernova. To achieve this, a considerable amount of energy (more than 10^{43} J for at least 10 seconds) has to be injected into them. This is equivalent to the energy emitted by our Sun during its whole lifetime – but compressed into just a few seconds. If the jets have insufficient force, the burst will be aborted, resulting in a classic gravitational supernova.

It appears that, while not all massive stars produce a burst, all bursts produce a supernova. The reasons why a massive star might produce a burst are still not clearly understood, and specialists in this field continue their lively and impassioned debates. Current studies do indicate that the history of these stars plays a decisive role, especially since, being massive, they exhibit very intense radiation pressure. One of the consequences of this pressure is the eventual expulsion of the hydrogen envelope before the death of the star. Now the extremely hot helium envelope is revealed to the observer, and the star has become a *blue giant*, a so-called Wolf-Rayet star (named after the two astronomers who first observed this type of object). The external appearance is therefore very different from that of a classic gravitational supernova, which, as already mentioned, appears in its late stage as a red giant.

Currently, it is estimated that less than 1 per cent of gravitational supernovae will produce a gamma-ray burst. Although there is much uncertainty about the figures, it is absolutely fascinating to contemplate the fact that, typically, one GRB explodes every minute in our universe. In order to

Figure 5.7 LISA (the Laser Interferometer Space Antenna) is designed to detect 'ripples' in the fabric of space-time called gravitational waves. Examples of events that cause them are massive black holes swallowing neutron stars or collisions between massive black holes. LISA will be the first mission to detect gravitational waves in space. To achieve the goal, three spacecraft form an equilateral triangle with arm's length of about 5 million km. Each spacecraft is kept centered around two cubes made of a gold-platinum alloy. The cubes float freely in space and the spacecraft protects them from the hazards of interplanetary space. The distance between the cubes in different spacecraft is monitored using highly accurate laser-based techniques. In this manner, it is possible to detect the minute changes caused by passing gravitational waves. (ESA.)

be visible, the jet projected from the resultant black hole must point in the direction of the observer. This explains the relatively small number of bursts actually detected.

Once the jet has been created and has left the envelope of the supernova, extremely violent mechanisms ensue, giving rise to the GRB phenomenon.

Thereafter, both long and short bursts will have similar histories, as we shall discuss below.

At the gates of Hell

As we have just seen, the progenitor for bursts of whatever type does indeed seem to be a newly formed black hole, surrounded by a belt of debris, the remains of a rapidly rotating massive star which has collapsed (long bursts) or the coalescence of two compact objects (short bursts). From this point onwards, the two categories will exhibit similar paths, which might lead us to infer some common physical phenomenon at their origin (Figure 5.8). The details of this phenomenon are the subject of particularly heated debate within the scientific

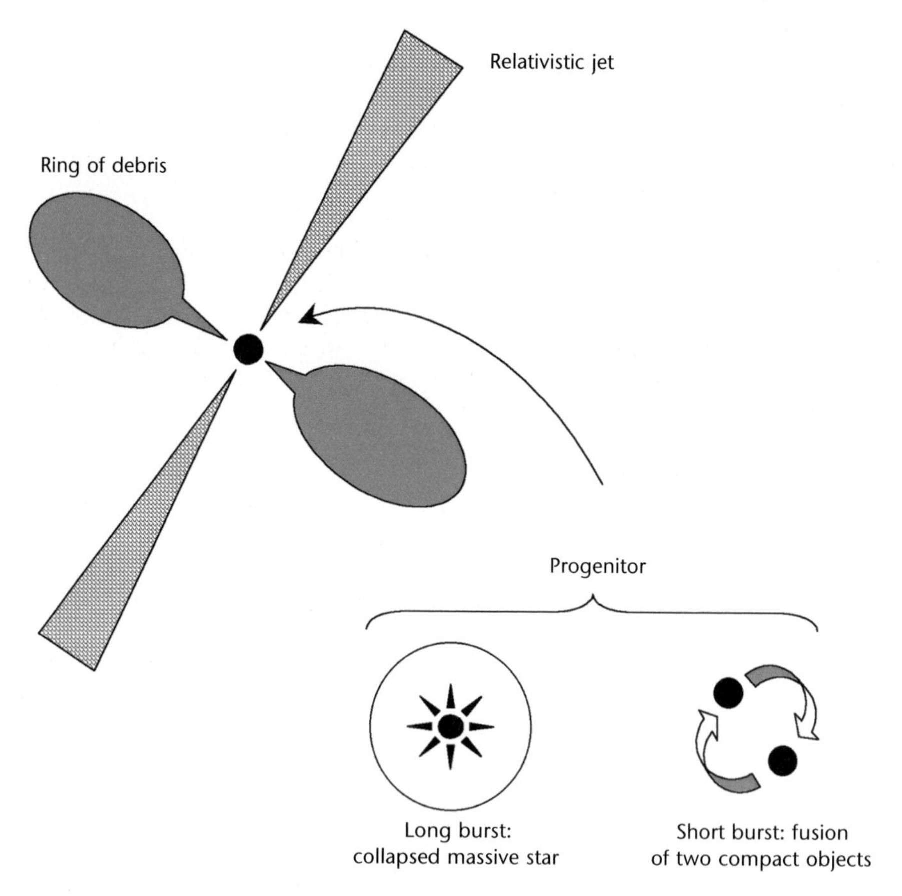

Figure 5.8 Diagram showing the principle of the formation of long and short gamma-ray bursts. The progenitor is always a black hole, newly created by the collapse of a massive star or the coalescence of two compact objects.

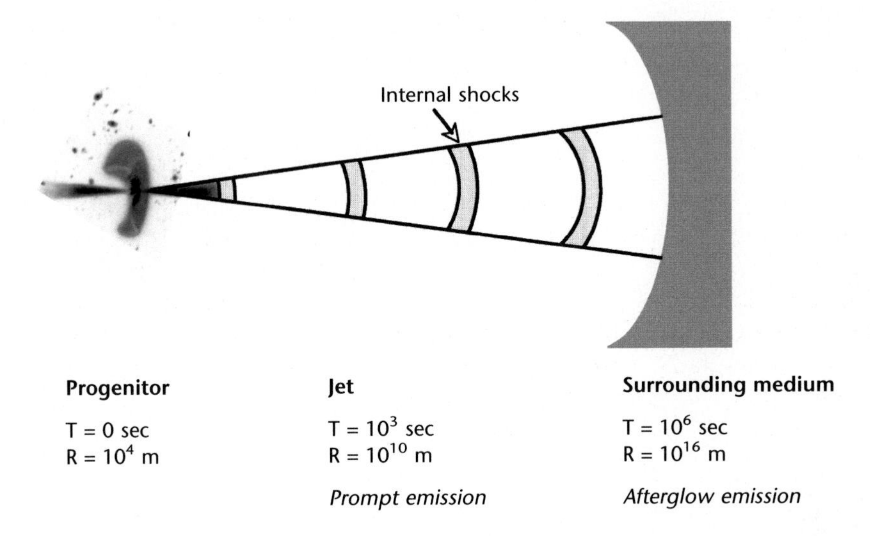

Figure 5.9 Diagram showing the principle of the 'fireball' model. Time and distance scales are shown in the frame of reference of the burst. The surrounding medium may be the interstellar medium, or matter previously expelled by the progenitor star.

community. Nevertheless, the so-called 'fireball' model gives insight into at least the broad picture. A fraction of the matter essentially composed of electrons will escape the gravity of the newly formed black hole, and these electrons will be ejected at *ultrarelativistic* speeds, in two exactly opposite cones. The opening angle of these cones is only a few degrees, and only by chance will the observer be situated in the direction in which they emerge.

A peculiar signature

The fireball is by no means homogeneous, and the electrons are ejected in an extremely chaotic fashion. Shells of electrons move out at very different velocities, and when one shell overhauls another, a more or less relativistic internal shock occurs, accelerating the electrons. These charged particles will radiate according to the *synchrotron effect* in the existing magnetic field, and the resulting emissions will represent a very wide section of the electromagnetic spectrum: from gamma-rays through X-rays to visible light. Several of these internal shocks may take place, causing the light curves of bursts to be very complex. This kind of emission is known as *prompt emission.*

As the fireball expands, it also sweeps away the material surrounding the progenitor. This may be material of the interstellar medium, or matter already ejected by the star (for example, its outer layers). This environment has a braking effect upon the fireball, and the associated electrons will also emit radiation across a very wide spectral range. This radiation is known as *afterglow emission.*

This description of a likely mechanism is naturally simplified; other phenomena, particularly involving fluid dynamics, are also present. In some

Figure 5.10 About 170 years ago, the southern star Eta Carinae mysteriously became the second brightest star in the night sky. In 20 years, after ejecting more mass than our Sun, Eta Carinae unexpectedly faded. This outburst appears to have created the Homunculus Nebula, pictured here in a composite image from the Hubble Space Telescope. Eta Carinae still undergoes unexpected outbursts, and its high mass and volatility make it a candidate to explode in a spectacular supernova sometime in the next few million years. (N. Smith, J. A. Morse (U. Colorado) et al., NASA.) See also PLATE 16 in the color section.

cases, a return shock wave propagates through the internal regions of the fireball following interaction with the external medium. In all cases, there are many collisions between different layers of matter; these cause the very great variability in the timescales of these phenomena. Each burst is different.

The overall scenario is shown in 5.9, with times and distances shown within the frame of reference of the progenitor.

Catching stars one by one

Gamma-ray bursts are, like their cousins the supernovae, absolutely extraordinary objects. They differ, yet, like human family members, they have their similarities. They outclass all others in the production of unimaginable energies. But the most fascinating thing about them is that they give us an insight into the universe across vast timescales reaching back through 95 per cent of its existence. Intrinsically luminous, and linked as they sometimes are to the deaths of the most massive stars, they offer astronomers a unique opportunity to observe individual stars out to the very edge of the cosmos.

We end this chapter on a note which, while apparently sombre, should not cause us to lose too much sleep. It would not be in our best interests for a gamma-ray burst to occur within our own Galaxy, and for its jet to be pointing our way. If this happened, the associated flux of X-rays and gamma-rays would be so strong that the ozone layer, our shield against the deadlier rays of the Sun, would be stripped away. Life on Earth would be under threat. Fortunately, we know of very few candidate stars. The nearest of these is Eta Carinae, about 8 000 light years away from us; but we are confident that the Earth is not in its 'line of fire' (Figure 5.10). However, the Galaxy is enormous, and we are familiar only with the stars in our locality.

The quotation from Racine with which we began this chapter may yet take on a somewhat broader meaning - but hope springs eternal....

6 Markers for cosmic surveys

'Nothing is easier than learning all the geometry; in case you ever need it'
Sacha Guitry

From the dreams of Hubble and Sandage, to dark energy

The cosmological 'supernova saga' really gathered momentum in the late 1990s, when two research teams, working independently on a sample of about 40 Type Ia thermonuclear supernovae, declared that, contrary to all expectations, the expansion of the universe is accelerating.[1]

This discovery has very important consequences for modern cosmology; consequences whose implications for physics itself have not yet been fully assimilated. It offers a coherent vision of our universe, giving rise to what has become known as the *Concordance Model.* This model gives us the opportunity to 'sort' the various pieces of the cosmic jigsaw puzzle which generations of astronomers have striven (with differing degrees of success) to provide. Observations of the cosmic microwave background over the years by (to name but a few) Boomerang, Archeops, COBE, and WMAP have led to an ever more accurate determination of the total amount of energy in the universe, expressed as the parameter Ω_{tot} (= 1). Studies of clusters of galaxies have led to an estimate of the quantity of matter in the universe: $\Omega_m = \sim 0.3$. Now, observations of supernovae have revealed a new contribution Ω_Λ to the total energy, its value being close to 0.7. We can see that these three pieces fit together very well indeed (Figure 6.1).

However, the most disconcerting conclusion that arose from these observations was that we now needed to imagine the existence of so-called 'dark' energy, dominating the universe and exhibiting a 'repulsive' gravitational action (the opposite of the usual attractive force inherent in matter, which causes Newton's apple to fall rather than fly upwards).

Such a concept is certainly surprising, but, if true, it could lead to a new Copernican revolution. What is not surprising is that very many efforts have been made, on the one hand, to confirm the measurements suggesting acceleration, and on the other, to try and identify the exact physical cause.

---

1. But... see *An Accelerating Universe* by J.E. Gunn and B.M. Tinsley in *Nature*, vol. 257, Oct. 9 1975, pp 454-457, where the idea was already being discussed.

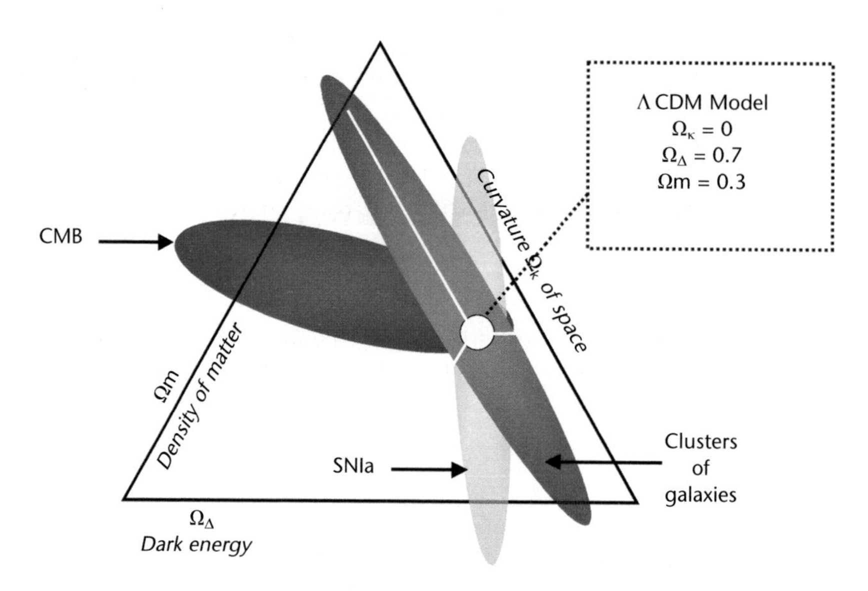

Figure 6.1 The cosmic triangle (or puzzle). It brings together the three key cosmic parameters: Ω_m, Ω_Λ, Ω_k. Each point of the triangle is such that $\Omega_m + \Omega_\Lambda + \Omega_k = 1$. The ellipses (with their associated uncertainty levels) represent the observational constraints based upon observations of the cosmic microwave background (CMB), clusters of galaxies, and Type Ia supernovae, with error margins. The parameters converge on the concordance model (or ΛCDM). Measurements of the CMB indicate flat space of zero curvature ($\Omega_{tot} = 1$, or the equivalent $\Omega_k = 0$), and clusters of galaxies a low density of matter ($\Omega_m \sim 0.3$). As for the Type Ia supernovae, they indicate an acceleration of the expansion and a new contribution to energy ($\Omega_\Lambda \sim 0.7$).

To better understand the issues involved, and their history, let us now take a step backwards.

Surveying the cosmos

From 1912 onwards, American astronomer Vesto Slipher, observing several dozen galaxies ('members of the Realm of the Nebulae') with a spectrograph, made the startling discovery that light from most of these objects was shifted towards the red. It was natural at the time to assume that this was a result of the Doppler effect. It meant that these galaxies were all receding from the observer, and suggested that the Milky Way lay at the centre of the universe – contrary to the Copernican Principle that there is no privileged direction or location in that universe.

As mentioned in the very first chapter, Edwin Hubble related the distances of various galaxies to their redshifts in 1929,[2] and from these measurements

2. Using the 100-inch (~2.5-metre) Mount Wilson telescope, at the time the most powerful in the world.

deduced his famous law, establishing that the universe is in an expansionary phase. In order to arrive at this result, Hubble drew upon the work of Henrietta Leavitt, who in 1908 had discovered a new class of variable stars, the Cepheids (named after the constellation of Cepheus, where the first example was discovered in 1784). The brightness of Cepheids varies due to regular phases of expansion and contraction. Leavitt showed that the period of variation in the brightness of Cepheids is related to their *absolute magnitudes* (known as a period-luminosity relationship), and can therefore be used to measure their distances. By measuring their *apparent magnitudes* (their brightness as perceived by an observer) and comparing it with their absolute magnitudes as determined by their period, it became possible to determine the distances of their host galaxies[3] (Figures 6.2 (a) and (b)).

A little (curved) geometry

It can be said that the establishment of Hubble's law, and its interpretation as evidence of an expanding universe, mark the moment when modern observational cosmology was born. No longer was the cosmos immutable, as humankind had imagined it to be for millennia: now, it had a history. A new era had begun, one of controversy and reversals of opinion. One notable such reversal occurred with Hubble's discovery (in collaboration with Milton Humason and Vesto Slipher) of the expansion of the universe: this lent support to the theories of Alexander Friedmann and Georges Lemaître, while Albert Einstein had stated that the universe could only be static. Einstein had to 'recant' the idea of the cosmological constant which he had introduced into his equations to ensure this static model of the universe. This constant (which Einstein later called 'the biggest mistake of my life') was to make a resounding 'come-back' 70 years on, in the context of the acceleration in the expansion of the universe.

The conceptual and mathematical framework of modern cosmology is based essentially upon the work of Einstein, and in particular on his theory of General Relativity, which gives insight into the effects of gravity on the grand scale. This cannot by itself suffice, however, and cosmology must also take into account certain postulates. The most important of these is the Cosmological Principle, which holds that space is homogeneous and isotropic, and more particularly allows us to define a *universal cosmic time*. This principle permits the definition of a chronology of events, and the writing of the history book of the cosmos.

Within this cosmic model, the universe is a space-time continuum, and generally curved (i.e. non-Euclidean), characterized by its dimensions and

---

3. Since the apparent magnitude decreases as the inverse square of the distance (i.e. $1/r^2$) in 'ordinary' Euclidean space.

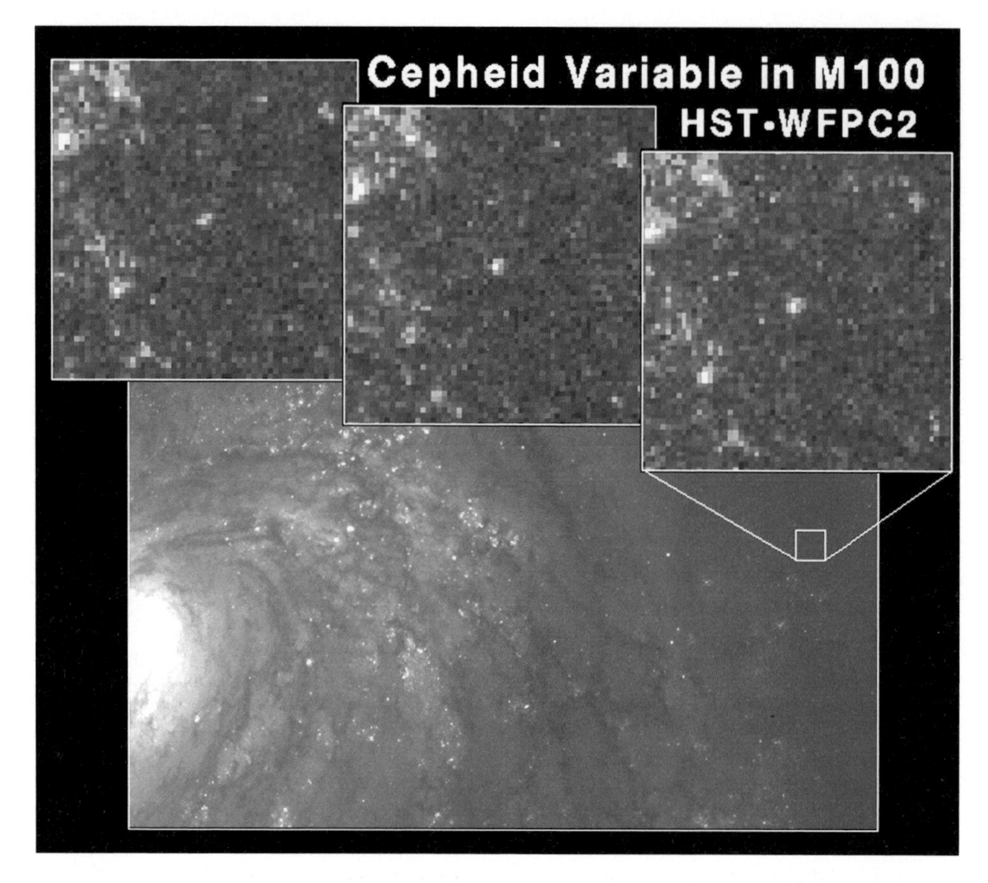

Figure 6.2 (a) This Hubble Space Telescope image of a region of the spiral galaxy M100 shows a pulsating star called a Cepheid variable. Though rare, these stars are reliable distance indicators to galaxies. Hubble pinpoints a Cepheid in a starbirth region in one of the galaxy's spiral arms (bottom frame). The top three frames, taken over several weeks, reveal the rhythmic changes in brightness of the star (in the center of each box). The interval it takes for the Cepheid to complete one pulsation is a direct indication of the star's intrinsic brightness. A longer cycle of brightening and fading means an intrinsically brighter star. By noting exactly how long the period of variation is and exactly how bright the star appears to be, one can tell the distance to the star and hence that of the star's parent galaxy – in the case of M100 it turns out to be 56 million light-years. Cosmic distance measurements as accurate as this are needed to calculate the rate at which the universe is expanding. (NASA, HST, W. Freedman (Observatories of the Carnegie Institution of Washington), R. Kennicutt (U. Arizona), J. Mould (ANU).)

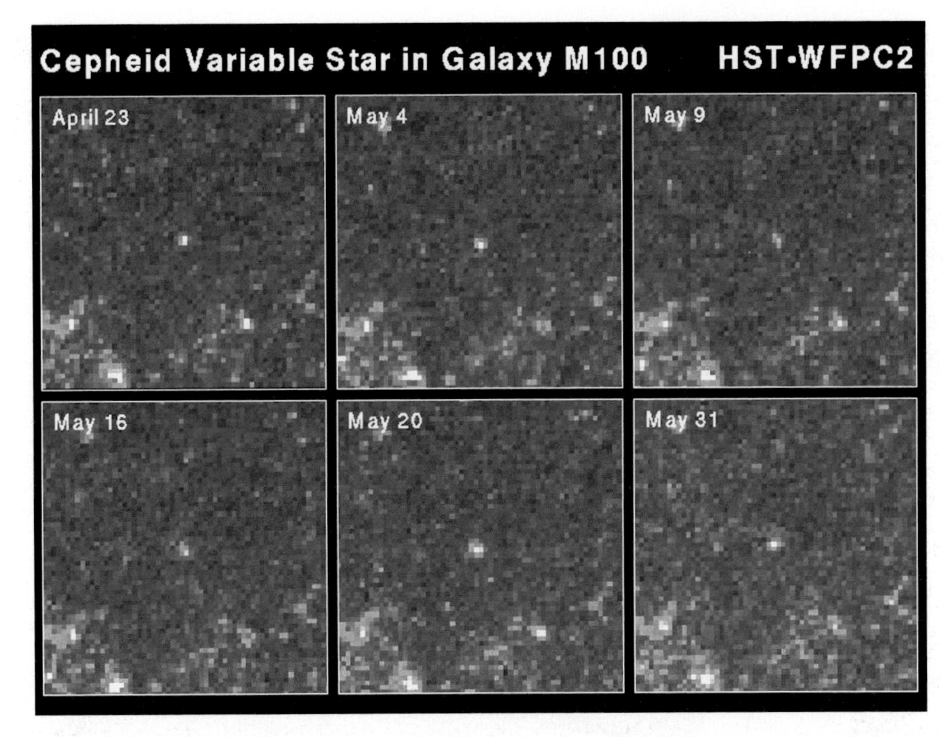

Figure 6.2 (b) This sequence of six Hubble Space Telescope images chronicles the rhythmic changes in a Cepheid variable star (located in the center of each image) in the galaxy M100. The Cepheid in these Hubble images doubled in brightness (24.5 to 25.3 apparent magnitude) over a period of 51.3 days. (NASA, HST, W. Freedman (Observatories of the Carnegie Institution of Washington), R. Kennicutt (U. Arizona), J. Mould (ANU).)

curvature (Figure 6.3). The properties of this space-time continuum are determined by its energy-matter content (and its nature), evolving over cosmic time. The two aspects, curvature and content, are related in Einstein's equations, which determine the behavior of the *scale factor R(t)* characterizing the expansion. It is not only the distances between the galaxies which depend on *R(t)*, but also the wavelengths of radiations and hence cosmic time.

The mean distance between the galaxies will therefore increase in an expanding universe, and this expansion may accelerate, decelerate, or even be reversed as a function of the energy-matter content (see Figure 6.4 and Appendix 10, page 141 for additional information).

The aim of observational cosmology is to determine this geometry and characterize the expansion by measuring the cosmological parameters involved. With this in mind, we can use the method of measuring the distances of ever more distant celestial bodies. Their behavior at various distances as a function of time will vary as the measured curvature, and the acceleration or deceleration of the expansion, vary. We can use our findings to differentiate between the

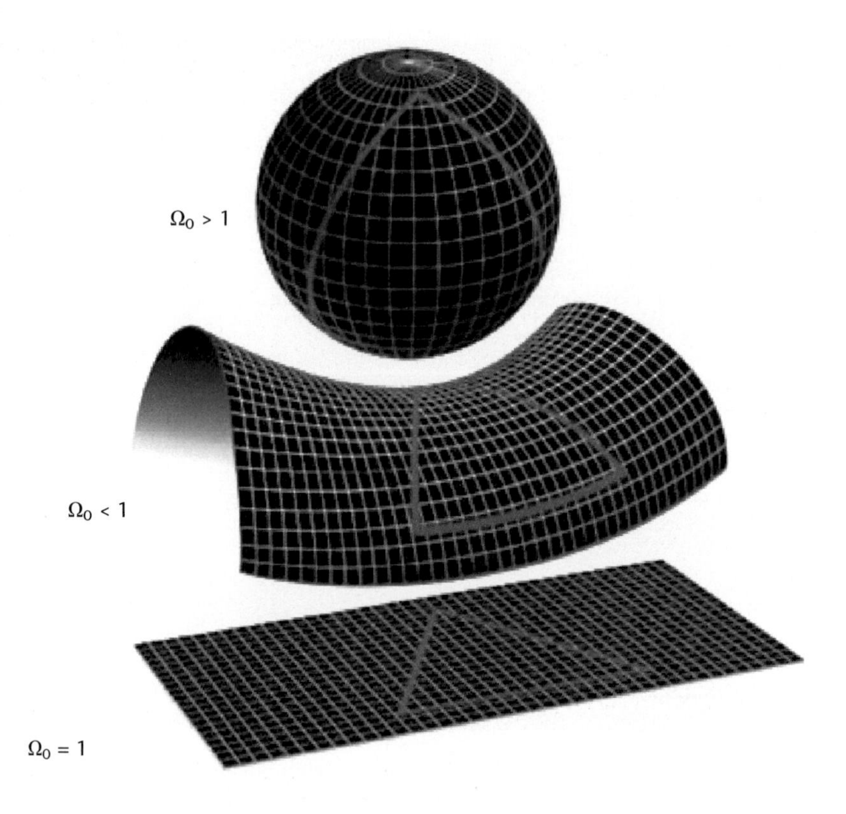

Figure 6.3 The universe is a four-dimensional space-time continuum (time, and the three usual spatial dimensions of width, length and height). Its geometry is characterized by curvature. It is illustrated here by two-dimensional surfaces. The curvature is created by the total energy-matter content of the universe as measured by Ω_0. As a result, we have, according to the value of this parameter, three types of geometry: spherical ($\Omega_0 > 1$), hyperbolic ($\Omega_0 < 1$), and flat ($\Omega_0 = 1$).

different cosmological models, and work out which are the best fit, and which do not fit. In curved space, several notions of distance are involved. If we are concentrating on the brightness of the stars, the pertinent parameter is distance-luminosity, traditionally written D_L (see Appendix 10, page 141).

So, if we can measure this type of distance, it becomes technically possible to compare the predictions of the various models with the observations, and to 'weigh' the different components of the universe.

A story of standards

As already mentioned, Hubble used a property of Cepheids, the relationship between their periods of variability and their absolute magnitudes, to determine the distances of the galaxies he observed. Measuring the Cepheids' apparent

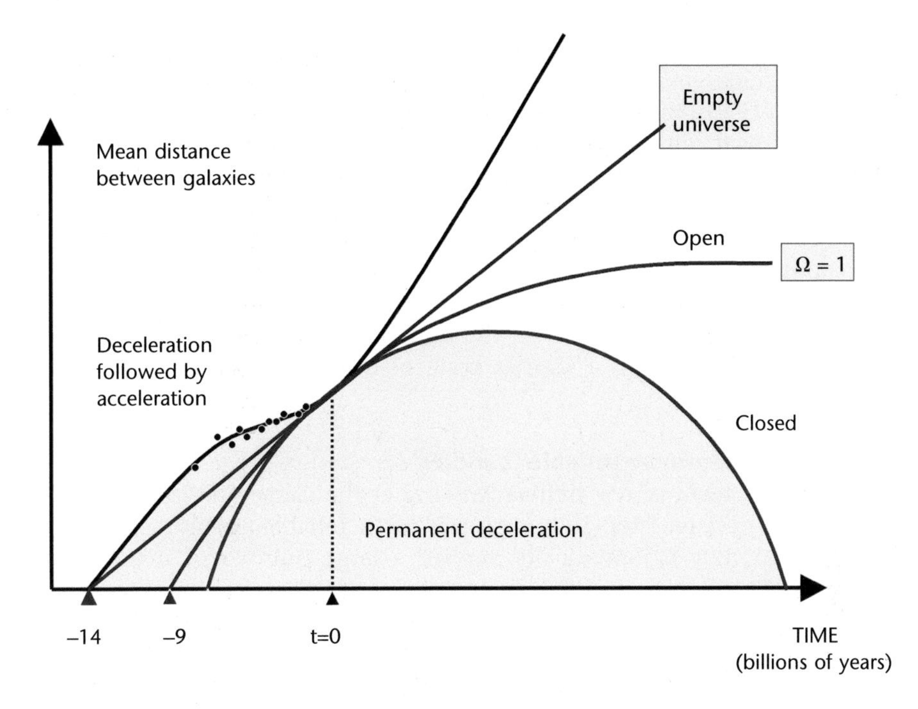

Figure 6.4 The history of cosmic expansion. The curves represent the evolution of the mean distances between galaxies, otherwise known as the scale factor $R(t)$, as a function of cosmic time. Depending upon the energy-matter content of the universe, the expansion is either continuously decelerated, or decelerated then accelerated. Measurements of the distances of thermonuclear supernovae (black dots) provide evidence of a phase of deceleration followed by a phase of recent acceleration.

magnitudes and periods allows the observer immediately to calculate their distance.

This principle can be extended to all categories of celestial bodies whose absolute magnitude is known: they can be used as *distance indicators*. The underlying hypothesis here is that objects of this type possess this property wherever they may lie in the universe and at whatever cosmic epoch we observe them. They are thus defined as *standard candles*.

To measure out the universe, both near and far, we need to use various families of such 'candles', intrinsically more and more luminous the further they are away. Cepheids can be intrinsically bright enough to be used out to distances of about 20 Mpc. Other 'candles' have been proposed, for example the brightest individual members of clusters of galaxies, thought to be of universally similar luminosity. In fact, such a standard may not be reliable, since the brightness of the galaxy in question may well have been modified over cosmic time by, for example, its merging with other galaxies. This is a phenomenon which goes on all the time and varies in frequency from one cluster to another.

Among other candidates, supernovae seem to be very promising 'skymarks', emitting as much energy as an entire galaxy. Their use in cosmology dates back to the 1960s–1970s, in the work of many pioneer researchers (including Colgate, Wagoner, Kowal, Branch and Patchett, and Kirshner) who were already anticipating the deployment of an instrument such as the Hubble Space Telescope (Figures 6.5 (a) and (b)). However, results were disappointing or even contradictory over the various scales of distance, given the difficulty of defining 'pure' standard candles, and disparities in the data obtained (often because of the use of too wide a variety of instruments/telescopes). This was one of the elements of the long drawn-out debate over the determination of the Hubble 'constant' H_0, with some proposing a shorter scale of distances and some proposing a longer scale.

Are Type Ia supernovae reliable 'candles'?

In the 1990s, a team of researchers working at the Cerro Tololo observatory in Chile took a decisive step in the search for a reliable cosmological distance marker, when they systematically studied a large number of nearby Type Ia supernovae. This first homogeneous sample made it possible for astronomers to study the properties of these supernovae and also their potential use in cosmology.

At distances such as those involved, the effects of the curvature of space are all but negligible, and the expected Hubble law is linear. However, it was found that the dispersion in the maximum luminosities of the supernovae (Figure 1.7) was still quite large,[4] which led to considerable uncertainty on the Hubble diagram. The *light curves* of these objects, though they are similar, are unfortunately not completely identical.

The decisive finding of these researchers was that the maximum in each light curve was linked to the manner in which the luminosity decreased as a function of time (i.e. the fall in the light curve): they had discovered how to link the difference in magnitude at its maximum and that observed fifteen days later. This was a key result, since by applying this correction, it became possible to greatly reduce the dispersion in the luminosities of the supernovae and, although they are not pure standard candles, they can be made so, if the corrections are applied. They are 'standardizable' candles.

Since then, other types of correction have been proposed in order to reduce the dispersion in intrinsic luminosity and improve standardization, by using other properties of supernovae. For example, there is the '*stretch*' technique, based on the fact that the light curves do not overlap each other (Figure 6.6 (a)), since the light curves of the brightest objects have a longer period than those of the less bright ones. We can therefore define a 'stretch parameter' (known as *s*)

---

4. The dispersion should be zero for absolutely pure standard candles, but, sadly, we live in an imperfect universe.

(a)

THE ASTROPHYSICAL JOURNAL, 232:404–408, 1979 September 1

SUPERNOVAE AS A STANDARD CANDLE FOR COSMOLOGY

STIRLING A. COLGATE
New Mexico Institute of Mining and Technology, and Los Alamos Scientific Laboratory
Received 1978 September 5; accepted 1979 March 9

ABSTRACT

Supernovae can perhaps be found at $Z \approx 1$ using the Space Telescope and the Focal Plane Camera (cryogenic charge coupled devices) at a rate of approximately four per week using 3 hours per week of viewing time. If Type II supernovae are used as a self-calibrating candle at $Z \ll 1$, then Type I's can be calibrated from Type II's as a secondary standard candle (2 mag brighter) and used instead of Type II's for a less difficult determination of q_0. This assumes all Type I's are the same independent of Z whereas each Type II is self-calibrated. Adequate statistics of supernovae in nearby galaxies $Z \lesssim 1$ can further verify the uniqueness of Type I's. Three-color wide-band photometry performed over the period of the maximum luminosity of a Type I gives the time dilation $\propto (1 + Z)^{-1}$, color shift $\propto (1 + Z)^{-1}$, and apparent luminosity $\propto Z^{-2}[1 + 0.5(1 + q_0)Z + O(Z)]^{-2}(1 + Z)^{-2}$. A Type I supernova at maximum and $Z = 1$, $H_0 = 50$, should give rise to a statistically meaningful maximum single pixel signal of ~ 250 photoelectrons compared to an average galaxy center background of ~ 25 photoelectrons for an 80 s integration time. An average of ~ 100 large galaxies ($10^{10}\ L_\odot$) per field allows $\sim 10^4$ galaxies to be monitored using 3 hours of viewing time. Z can be determined by time dilation and color shift sufficiently accurately that the determination of q_0 will have twice the error of the calibration of Type I as a standard candle.

Subject headings: cosmology — stars: supernovae

(b)

Figure 6.5 (a) In the late 1970s, it was recognized that observing supernovae, allied to the use of space instruments such as the (future) Hubble Space Telescope, would offer exceptional investigative insights into cosmology 'à la Sandage'. (*The Astrophysical Journal.*) **(b)** Edwin Hubble's name was given to the space telescope, eventually launched in April 1990, whose rich harvest would revolutionize astrophysics in all its aspects. (NASA). See also PLATE 18 in the color section.

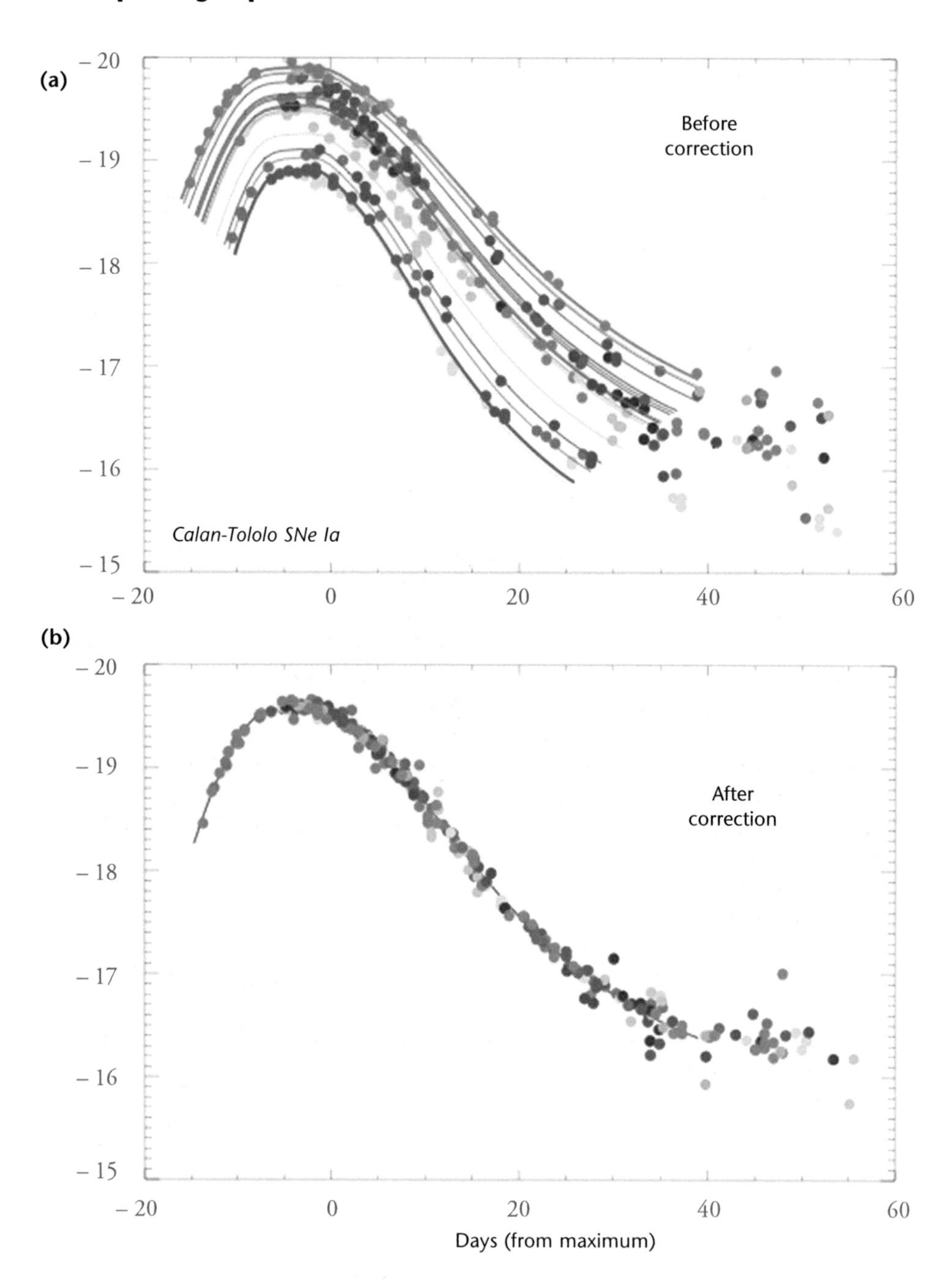

Figure 6.6 (a) Light curves of Type Ia supernovae do not intersect, but can be effectively superimposed once the *stretch* method is applied, as shown in **(b)**.

which characterizes the duration of the decline of the supernova: a parameter used to adjust all the light curves to the same size[5] (Figure 6.6 (b)).

The hunt for Type Ia supernovae

In any given galaxy, supernovae are rare objects, at once random and fleeting. In order to be of use in cosmology, a sufficiently large sample has to be observed across a considerable range of distances (i.e. volume of space) if we are going to be able to differentiate effectively between the various models of the universe. Also, the supernovae must be observed over a period of several months, in order accurately to study the properties of their light curves.

It may not be difficult to detect a new supernova – *if* you have the patience. The fact is that supernovae occur frequently on the scale of the whole cosmos, with its billions of galaxies. If we constantly observe an area of the sky, we are certain to find a supernova sooner or later.

So the strategy is quite a simple one, consisting of obtaining a 'deep' image of a given area of the sky, which will serve as a reference. Later, we take another image of the same area and compare the two. Among the variable objects detected, a fraction[6] will certainly be supernovae (Figure 6.7).

The other necessary ingredient for cosmological study is the accurate measurement of the redshift z of the supernova. This is obtained spectroscopically with very large telescopes (8 to 10 meters in diameter), for example, the VLT (Very Large Telescope) in Chile, the Gemini telescopes in Chile and Hawaii, and the Keck telescopes in Hawaii (Figures 6.8 (a)–(f)).

This is all very well in (observational) theory, but in practice it is not quite as simple. For the most distant supernovae we need not only a large telescope capable of detecting faint objects by collecting as many photons as possible, but also a detector capable of covering a very wide field[7] in order to stand the best chance of observing a supernova.

Such instruments are unfortunately rather rare, which is why only a few research programs have been initiated on telescopes in the 4-metre class, such as the Canada-France-Hawaii Telescope in Hawaii (Figure 6.8 (a)) and the Blanco Telescope of the Calan-Tololo Survey (Chile). These telescopes and their associated high-quality instruments offer a unique opportunity of covering a large area of the sky (between 0.5 and 1 square degree) in one exposure, accessing volumes of the universe large enough to be used for systematic research.

---

5. Another correlation, known as 'brighter-bluer', has been put forward and allows even greater reduction of the dispersion of luminosity to a maximum L_{Max}. This establishes the relationship between L_{Max} and the colour of the supernova.
6. There are other variable objects in the universe, for example, novae and quasars,
7. In order to search efficiently for supernovae, the detector must cover an area of at least half a square degree, roughly equivalent to the size of the Full Moon in the sky.

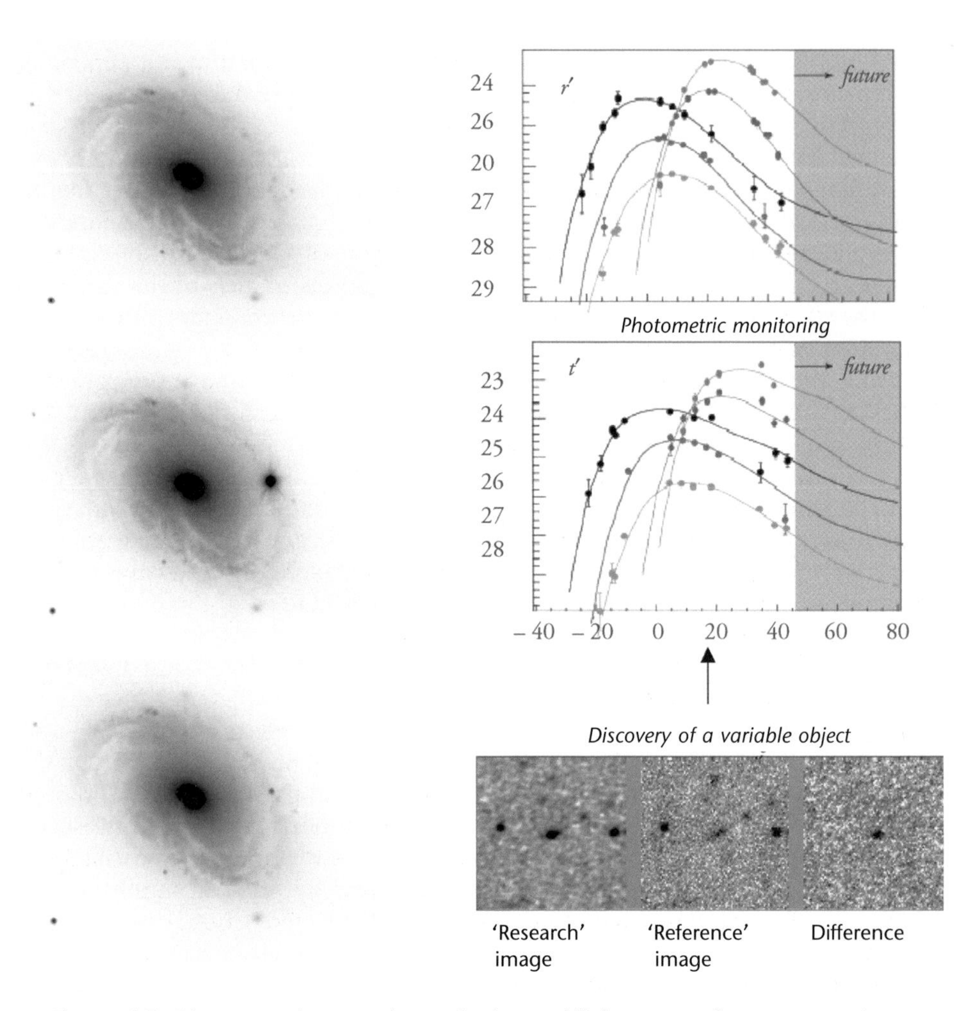

Figure 6.7 Discovery, photometric monitoring and light curves of supernovae. A research image is compared to a reference image previously obtained. The difference between the two images will reveal a certain number of variable objects, including *supernovae*. Subsequent monitoring of the chosen candidate stars is carried out using filters of different wavelengths (here, *i'* and *r'*) to bring out the properties of these objects.

'Accurate cosmology'

Convergence towards Concordance?

As we have already suggested, current projects aim not only to 'test' the expansion and its acceleration, but also to try and identify the cause. These goals, and particularly the second, require measurements of very great accuracy. First, the pieces must slot together ever more perfectly in the 'cosmic triangle', which necessitates the determination of the three basic 'cosmological parameters' (Ω_{tot}, Ω_m, Ω_Λ), to an accuracy of 99 per cent or even greater; and also it is hoped to distinguish, for example, a 'true cosmological constant' of 'time-dependent dark energy'.

A very good example of progress made in recent years can be found in the results of the Supernovae Legacy Survey (SNLS). Currently, this program has compiled the largest homogeneous sample of Type Ia supernovae. With results for more than a hundred distant supernovae associated with data obtained elsewhere for nearby supernovae, a first analysis (Figure 6.9) shows that a cosmological model *without* a contribution from a cosmological constant or from dark energy is totally out of the question. The model of a universe containing nothing but matter lies systematically outside the error margins of the measurements. Other programs such as that utilising the Hubble Space Telescope to detect and observe much more distant Type Ia supernovae, are pointing us in the same direction.

However, these results are still not definitely conclusive. What is needed is to demonstrate in no uncertain terms that this is indeed a real effect and not due to some other phenomenon mimicking the expansion of the universe. Therein lies the whole challenge.

The cosmos is subtle...[8]

Different effects can distort, or even invalidate our interpretation of the variation in the brightness of supernovae in terms of acceleration or expansion. One of the main difficulties is that the properties of supernovae may have been modified during the history of the universe. Of course, the most distant examples of these bodies exploded several billion years ago, in galaxies which have since harbored several generations of stars. The environment, or the progenitor stars themselves, of the Type Ia supernovae may have varied, if only slightly, from one epoch to another.

One method of evaluating the effects of evolution is to compare the properties of supernovae at greater and lesser distances, i.e. in different cosmic eras. Despite numerous studies, no statistically significant difference has yet been revealed. One of the best proofs is furnished by SNLS analyses, comparing the main properties of supernovae discovered in the course of the program with those of nearby supernovae. If there had been any evolution over cosmic time, the two distributions would exhibit statistical differences, as might be expected for example in a survey of the height of a human population as a function of age. Yet an analysis of these data indicates no noticeable evolution.

---

8. ... *but it is not malicious* – to paraphrase Einstein.

Figure 6.8 (a) To observe distant supernovae, we need to use telescopes of at least 4 meters aperture, such as the CFHT (Canada-France-Hawaii Telescope) in Hawaii, shown here at night with star trails behind, both to discover them and to monitor their light curves. (Jean-Charles Cuillandre (CFHT).) See also PLATE 17(a) in the color section.

(b) Associated with these telescopes are wide-field cameras, such as MegaCam, which covers about 1 square degree of sky, and is at present one of the largest imagers in the world. (Paris Supernova Cosmology Group.) See also PLATE 17(b) in the color section. Identification and distance measurements of supernovae are carried out spectroscopically by the largest telescopes (8–10 meters) such as:

(c) the unit telescopes of the Very Large Telescope (VLT) in Chile (ESO)

(d) the Gemini telescopes in Chile and in Hawaii (Gemini Observatory/Keith Raybould.), and

(e) the Keck telescopes in Hawaii. (W.M. Keck Observatory.) See also PLATE 19 in the color section.

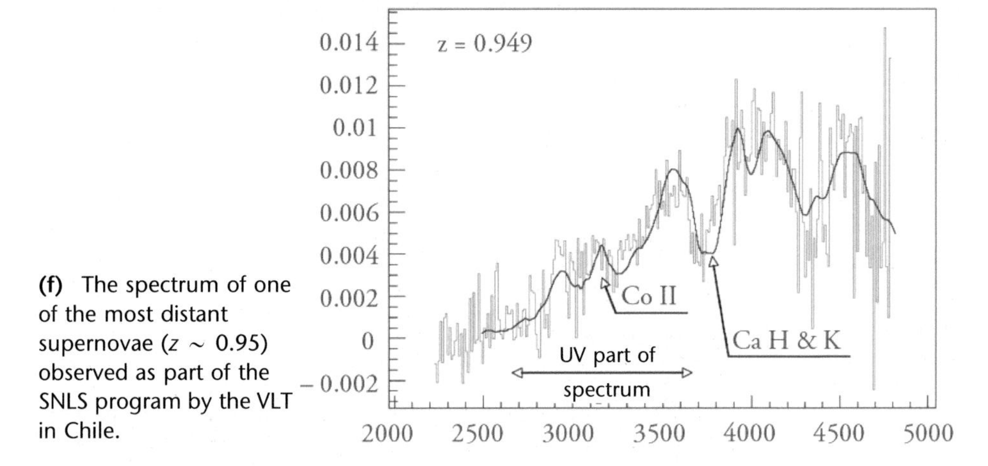

(f) The spectrum of one of the most distant supernovae ($z \sim 0.95$) observed as part of the SNLS program by the VLT in Chile.

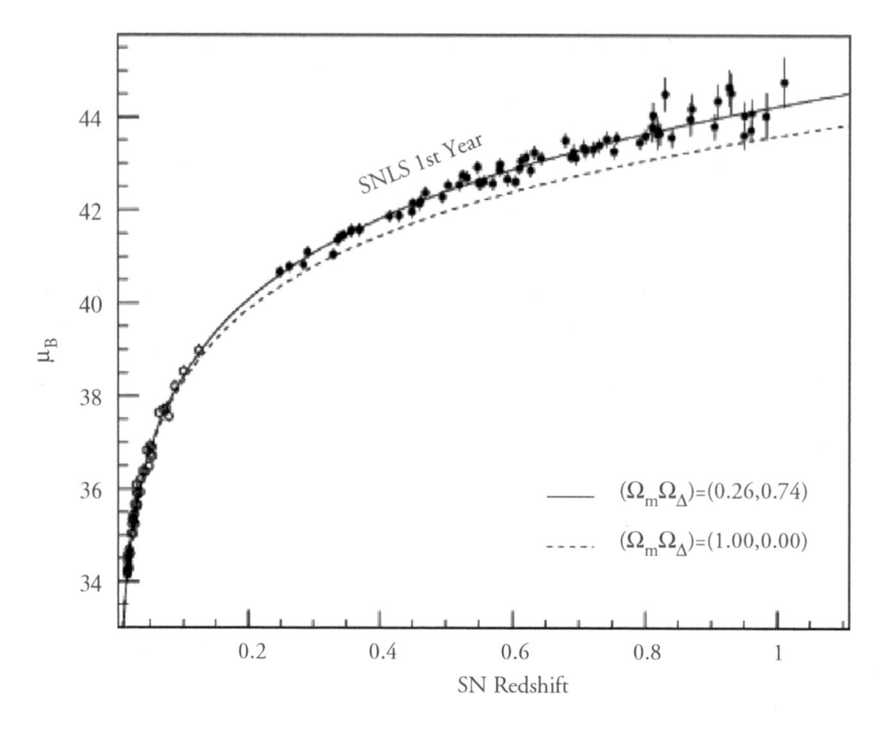

Figure 6.9 Example of a Hubble diagram obtained using observations made during the first year of the SNLS program. The positions of Type Ia supernovae are indicated by the small circles; predictions are shown for a universe with (solid line) and without (dashed line) dark energy. Clearly, a universe without this new component would not tend to explain our observations!

Even though this question of the evolution of supernovae is still a subject of keen debate within the scientific community, no variation (within the context of the margins of error of the measurements) is clearly identifiable. This indicates, at least on the cosmic timescale involved, that evolution does not seem to be a factor, and will therefore not affect the original conclusions.

Other troublesome effects can also intervene to attenuate the brightness of supernovae, for example, the presence and action of dust within their host galaxies, or along the path of the emitted light.

During recent decades the extinction due to intra- and extra-galactic dust has been systematically mapped by space missions working in the infrared, such as the ISO and COBE satellites. These observations can be used to apply the necessary corrections in the context of our own galaxy.

However, it is known that the quantity of dust within the host galaxies of the supernovae was certainly greater in the past, an effect which can easily slant the conclusions drawn from the Hubble diagram, by lessening the brightness of the supernovae. In order to investigate the influence of this dust, we elaborate so-called 'reasonable' models of the nature and distribution of galactic dust. These models are then tested by comparing their predictions with the observations, and particularly their influence on the observed colors of supernovae. The presence of dust tends to 'redden' the perceived radiation since the dust preferentially removes shorter wavelength photons while leaving behind the longer wavelength photons - in the optical, light that is redder. In every case, no significant effect has been detected, within the limits of current uncertainties in the measurements. The same is true in studies of other objects: for example, distant quasars.

Another particularly important and very frequent effect in astronomy is known as the *Malmquist bias*. This refers to the fact that, the nearer we approach to the limit of detection of our telescopes/detection systems, the more we tend to select, for a given distance, the brightest objects, unduly favoring them. Analysis of actual data indicates that these data do not seem to be systematically biased[9] because of this effect.

In practice, we add a contribution corresponding to an estimate of this effect to the global 'error budget' of the final accuracy which we give to the measured parameters.

One last 'subtle' effect is related to the propagation of light in a universe where the distribution of matter exhibits heterogeneities: the large-scale structures, such as galaxies, and clusters and filaments of galaxies. The presence of massive objects causes light rays to deviate: this is the well known gravitational lensing effect, which can influence the brightness of bodies observed. On average, distant supernovae are dimmed, an effect which could invalidate cosmological analyses. Also, to evaluate the expected effects, we base our estimates (and they are indeed tenuous) on realistic models of the distribution of the large structures in the universe, and take account of them in the global error budget.

Obviously, all these tests may seem superfluous and tedious, but they are really necessary, and are the price we pay to assure the reliability of our scientific results. This approach, then, is the very bedrock of the scientific method, which consists in the constant questioning of the results obtained. It is indispensable to demonstrate that supernovae are dimmed because of the accelerated expansion of the universe, and not by, for example, their temporal evolution or the effects of attenuation by dust.

If we are indeed on the verge of a new Copernican revolution, we had better be sure of our propositions.

From supernovae to gamma-ray bursts

And further still...

As we have just seen, the use of standard candles is an extremely powerful tool for our survey of the universe. By using Type Ia supernovae, it has been possible

---

9. Also, we observe more objects which are 'too bright' (such as supernova SN 1991T) in samples of the nearby universe than in more distant samples, which runs counter to the expected effect.

to demonstrate, and convincingly for the first time, an 'accelerating universe'. This remarkable result, since confirmed by indirect measurements, is one of the key discoveries of recent times, since it has given rise to the now famous concept of dark energy.

Even though their intrinsic luminosity is much greater than that of variable stars, this particular class of supernovae is limited to redshifts of about 2 (representing a period of time of about 8 billion years). To gauge the expansion of the universe even more finely, it is therefore necessary to make use of new 'standard candles', whose range in terms of luminosity will be greater, and allow us to survey the universe over even longer timescales. We must still, of course, take certain effects, such as absorption by dust and the Malmquist bias, into account.

So it is very tempting to try to use the most energetic bodies in the cosmos: gamma-ray bursts. These offer two remarkable properties: they are extremely luminous, and might be visible out to redshifts of 10-15, i.e. to a time when our universe was less than 500 million years old. The fundamental question is whether these gamma-ray bursts can be considered as standard candles, or at least as standardizable candles, as is the case with supernovae.

Standards to be confirmed

First, it is important to recall some of the properties of gamma-ray bursts in order to be able to pursue this story. These bursts are, by definition, very brief and transitory phenomena (lasting from a few seconds to a few minutes), emitting over a very wide spectral range, from gamma-rays to visible light.

Two classes are traditionally identified, as a function of their duration: short bursts and long bursts. In the case of the latter, the most commonly accepted model is at present based upon the core collapse of an extremely massive star (more than 20-30 times the mass of our Sun) and the formation of a central black hole. This process leads to the formation of a narrow jet of matter, travelling at near-light speed.[10]

However, the different layers of matter within the jet do not propagate at exactly the same speed as each other, creating *internal shock waves*. The energy liberated by the shocks is responsible for the observed 'prompt' emission. The jet continues its propagation, colliding with the external layers of the star, and then with the interstellar medium, creating new shocks, and slowing as it broadens out. A subsequent ('afterglow') emission results from the interaction of the jet with this immediate environment.

By definition, a 'candle' in cosmology is an object of universal absolute energy. In the case of a long burst, the progenitor is a massive star, whose initial mass is unknown and within which poorly understood phenomena occur. So are they a lost cause?

---

10. The term used is 'relativistic jet', emitted according to the Lorentz factor (see Appendix 8, page 137).

Mother Nature can sometimes be kind to cosmologists.

At first sight, and as if it were a 'classic' astrophysical situation, we might be tempted to deduce the total energy from the number of photons received, supposing that they have been emitted isotropically (energy usually denoted by $E_{\gamma,\ iso}$). If we consider the distribution of this quantity in the hope of finding the possibility of a standard energy, then we shall be somewhat disappointed.

However, to proceed thus would be to ignore the fact that the energy of a gamma-ray burst is emitted in collimated jets, whose opening angle we do not (*a priori*) know. One thing in our favor is that the mechanisms operating in the jets are relativistic and 'relatively' easy to model. In effect, the jet slows down as it encounters matter during propagation into the surrounding medium. Its (near-relativistic) speed will eventually fall, at a time t_c, below a critical value, and this deceleration will cause a considerable spreading of the jet in question. An abrupt change in the shape of the light curve (the *jet break*) will occur (Figures 6.10 and 6.11).

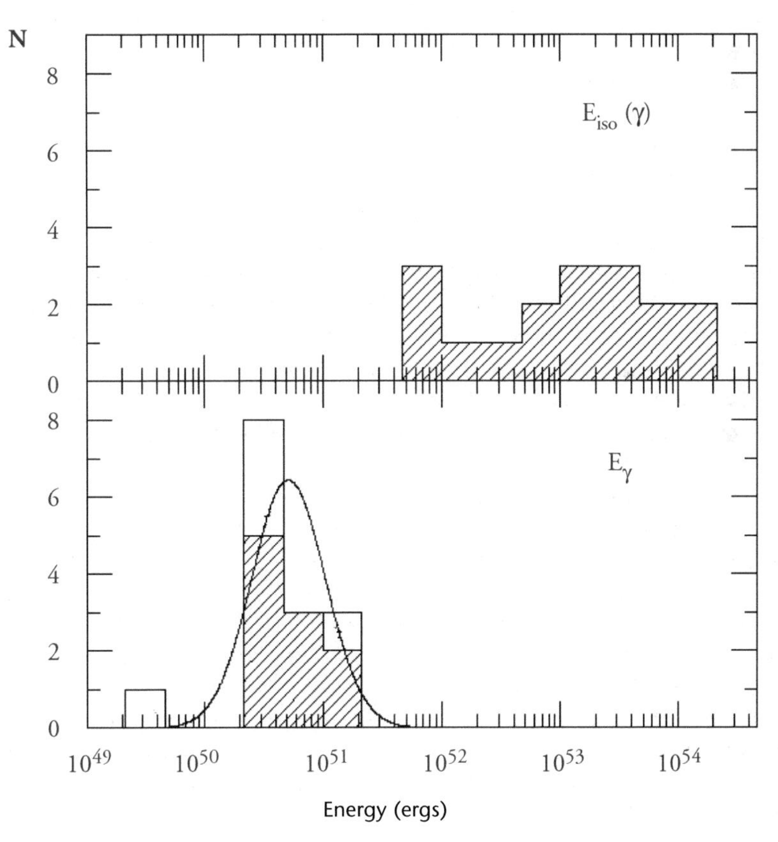

Figure 6.10 Distribution of 'isotropic' energies E_{iso} (γ) and energies corrected for 'beam effects', showing that bursts may also be 'standard candles'.

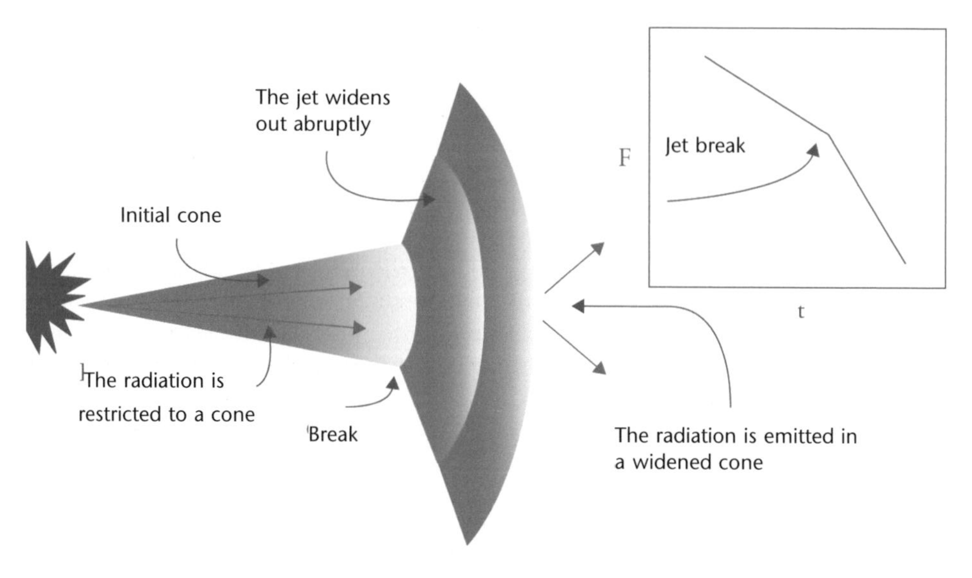

Figure 6.11 As it propagates, the relativistic jet changes its nature, at a moment in time t_c. This slowing causes an abrupt widening of the jet, and an observable change in the direction of the light curve. It is then theoretically possible to relate the time t_c to the opening angle θ of the cone.

Now, according to simple but reliable models, the time t_c, which can be determined by observation, is related to the opening angle θ of the jet. It therefore becomes possible to calculate the energy: not in terms of an isotropic value, but of that (E_γ) physically contained within the jet.

We then see that this energy is, contrary to the previous case, grouped in all instances around the same value, of the order of 10^{44} joules (10^{51} ergs), with a very small dispersion (lower panel of Figure 6.10). Thus, the bursts achieve the status of 'possible standards'. But the dispersion is still too large, and, as in the case of the supernovae, we seek to determine if there might not be a way of reducing it. Mother Nature smiles upon us again. We know that the quantity E_{Peak}, characteristic of the spectra of bursts, is related to the total energy corrected for the effects of collimation (E_γ). Just as we used the *stretch* technique to align the light curves of the supernovae, we can use the relationship between the energy E_γ and the quantity E_{Peak} to standardize the bursts.

And there you have it. Long gamma-ray bursts seem well and truly standardizable.

A standard marriage

Armed with this new standard, we are naturally tempted to marry together observations of Type Ia supernovae and long gamma-ray bursts. There is indeed a good agreement between these different measurements. Once more, the Concordance Model is confirmed with a universe currently in a phase of accelerated expansion, a universe whose energy-matter content is dominated by dark energy.

The result is extremely encouraging, and underlines how productive the combination of data furnished by various probes can be. So the gamma-ray bursts do not exhibit the same bias as the supernovae. For example, gamma radiation is only weakly affected by dust, and the bursts are therefore not affected by extinction due to dust, as may be the case with supernovae.

We must remain careful, though. Uncertainties linger about the use of bursts, because there are not enough homogeneous data. Currently, the data are fairly disparate and the number of objects for which complete data are available is small. For example, one essential piece of information, the distance of the burst, is known for fewer than 100 objects, even though several thousand have been detected. The danger lies therefore in accidentally choosing a class of objects that mimic the effect which one is seeking to observe.

Moreover, in the case of the supernovae, we know that there is an underlying universal quantity (the Chandrasekhar mass) involved, even if we do not know in detail the mechanism at the origin of their brightness; but in the case of the bursts, we know nothing of the sort. Even if we thought that the mechanism by which these objects emit radiation were 'universalized', while discounting the details of their progenitors, the debate on which is the best technique to use is still raging within the scientific community.

The use of gamma-ray bursts to survey the universe is still a very young enterprise, but it holds great promise and there is much to be done, both observationally and theoretically. The situation is in some ways similar to that before 1998, with the study of Type Ia supernovae, prior to the unveiling of dark energy.

As we shall see in the final chapter, the systematic observation of gamma-ray bursts will be a major objective during the next decade: for coming space missions, for high-energy astrophysics and for cosmology.

7 Beacons in the cosmos

'That which illuminates always remains in darkness'

Edgar Morin

Prestigious precursors

Troublesome black holes

During the 1950s, many 'point' radio sources, i.e. sources similar to stars and emitting in the radio domain, were detected in the sky, and had been catalogued without any optical counterpart having been identified. Telescopic observations of their locations revealed nothing. Then, in 1960, the radio source 3C48 was found to coincide with a blue, star-like object. However, the spectrum obtained resembled nothing previously seen. Unlike the spectra of stars, which exhibit many absorption lines, this spectrum showed very wide emission lines.

About two years later, another such object - 3C273 - was linked to an optical source with similarly odd spectral characteristics. Astrophysicist Maarten Schmidt came to realize that these represented redshifted hydrogen lines, and opened a new window upon the cosmos (Figure 7.1 (a)).

Just as the Rosetta Stone had opened the way to the decipherment of hieroglyphs, this observation led to the interpretation of the spectra of other sources, and a velocity of recession of about 100 000 kilometers per second was ascribed to 3C48. These objects, which looked like stars but were radio sources, were given a name by astrophysicist Hong-Yee Chiu: *quasars*, from 'quasi-stellar radio sources'. They are also known as QSOs ('quasi-stellar objects'). 3C 273 is the optically-brightest quasar in our sky, and one of the closest with a redshift of $z = 0.158$, corresponding to a distance of about 600 Mpc (Figure 7.2).

These newcomers in the cosmic menagerie became the objects of a frenetic search and of intense discussion. One debate revolved around their 'true nature' in the cosmos, because the energy emitted by such distant objects implied extremely powerful mechanisms. Once their cosmological nature had been recognized, discussion turned to the identification of the mechanism(s) responsible. As more and more of them were discovered, the complexity of their observed properties challenged the reliability of the contemporary model of the universe.

Since that time, doubts have vanished, and the cosmological nature of these objects is firmly established. As for the mechanisms responsible for their formidable energy output, and for that of their fellow 'newcomers' in the cosmos

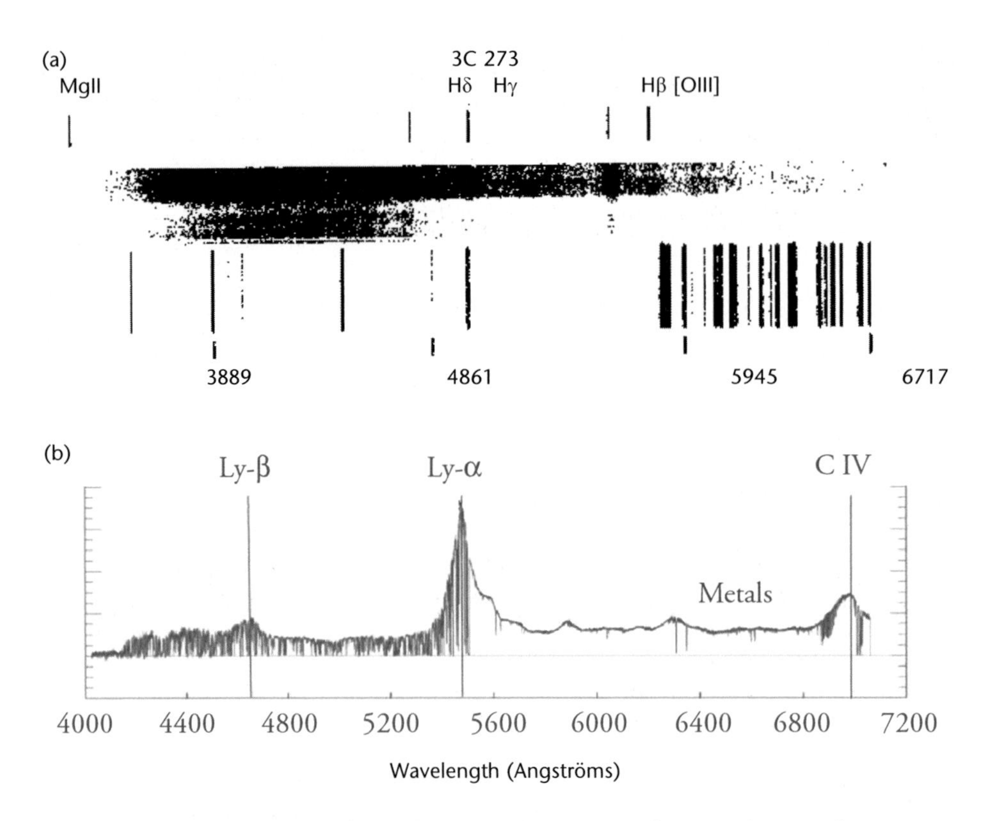

Figure 7.1 (a) The original two-dimensional spectrum of 3C273 showing the hydrogen lines which revealed its extragalactic nature.

(b) A recent one-dimensional spectrum of 3C273 showing the Lyman-α emission line, indicating a redshift $z \sim 0.16$. Note the extraordinary improvement in quality between these two measurements, due to the enormous progress made in the design of telescopes and other astronomical instruments.

(blazars, radio galaxies etc.), the factor differentiating the various kinds of active galaxies seems to be only the angle from which we observe them, concealing or revealing diverse aspects such as jets and discs. They now fit into a coherent description of a model involving an accretion disc orbiting a supermassive[1] black hole which is emitting a relativistic jet.

It is the accretion by the black hole of enormous quantities of matter that is responsible for the vast energy emitted by these sources.

A forest full of treasures

As well as exhibiting the wide emission lines (such as the Lyman-α in Figure 7.1 (b))

---

1. i.e. between 10^5 and 10^{10} times the mass of the Sun.

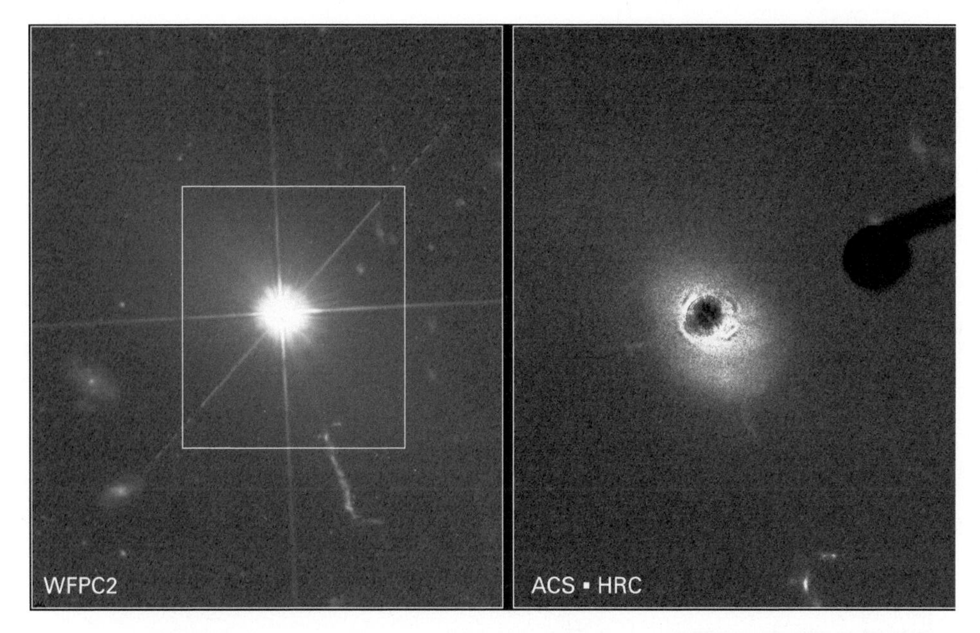

Figure 7.2 Left: This image of the quasar 3C273 in the constellation of Virgo, acquired by the Hubble Space Telescope's Wide Field Camera 2, shows the brilliant quasar but little else. The diffraction spikes demonstrate the quasar is truly a point-source of light (like a star) because the 'central engine' is so compact; the quasar contains a supermassive black hole about one billion times the mass of the Sun. (NASA and J. Bahcall (IAS).) **Right:** Once the blinding light from the brilliant central quasar is blocked by the coronagraph of Hubble's Advanced Camera for Surveys (ACS), the host galaxy pops into view. Note that the ACS' occulting 'finger' and other coronagraphic spot are seen in black to the right of the ACS High Resolution Channel image. (NASA, A. Martel (JHU), H. Ford (JHU), M. Clampin (STScI), G. Hartig (STScI), G. Illingworth (UCO/Lick Observatory), the ACS Science Team and ESA.) See also PLATE 20 in the color section.

that are the signature of very energetic phenomena in action, the spectra of quasars characteristically show a compact series of absorption lines known as the *Lyman-alpha forest.* This crowded set of lines results from the absorption of the radiation emitted by the quasar (Figure 7.3) by all the matter lying along the line of sight between the source (the quasar) and the observer. Obviously, radiation will be absorbed by matter in the immediate vicinity of the quasar, but it is also absorbed by matter (essentially hydrogen) within the vast intergalactic clouds left over after the formation of large-scale structures.

So this 'forest' gives us a considerable amount of information about those large structures, out to very great distances, and the study of the characteristics of the Lyman-α forest constitutes a further test of their formation scenarios (Figure 7.4). Also, we are able to measure the quantity of baryons associated with the structures responsible for the Lyman-α forest, for all redshifts (i.e. as a function of cosmic time), and compare it with the predictions of primordial nucleosynthesis.

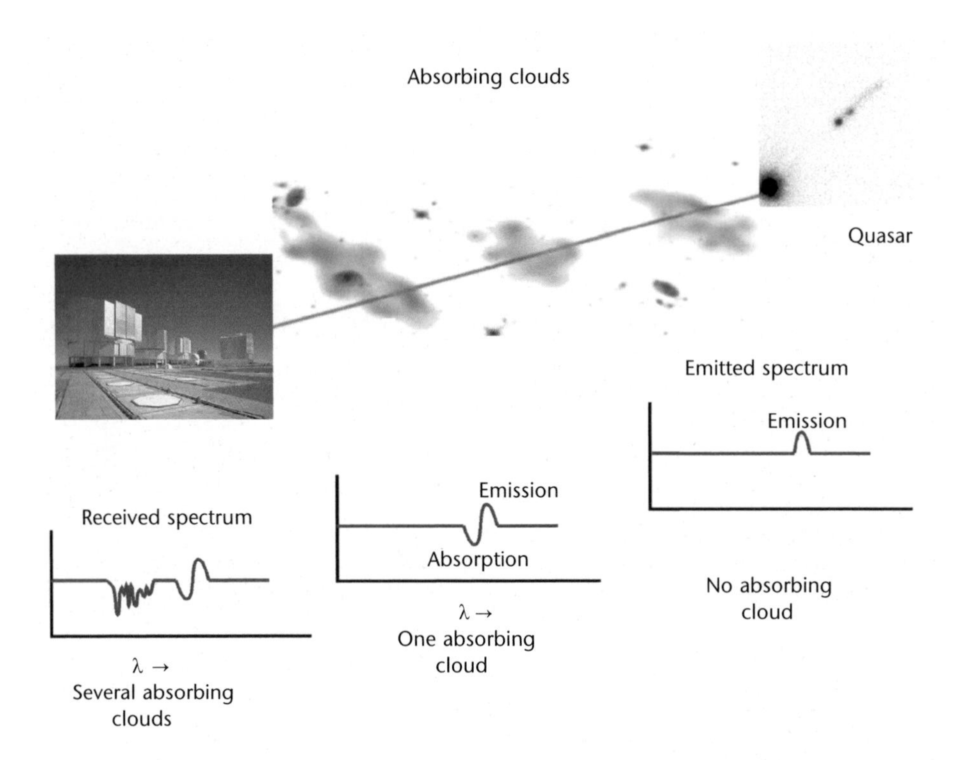

Figure 7.3 The quasar emits radiation redshifted by the quantity (1 + z), as is the Lyman-α line at approximately 5 500 Angströms in Figure 7.2 (b). Each hydrogen atom (for which Lyman-α is the fundamental line), situated at a smaller redshift than that of the quasar in question (and therefore in front of it from the observer's point of view), can absorb photons at all bluer wavelengths. The totality of the absorptions all along the line of sight creates the 'forest' effect.

Gamma-ray bursts to the rescue

Since gamma-ray bursts are even more energetic than their bigger siblings the quasars, it seems very likely that they can also be similarly useful in our searches of the cosmos, and provide information about regions even further away in both distance and time.

Cosmic Pop

Since GRBs are stellar in nature and observable at potentially (very) great redshifts, they seem to be ideal candidates to advance research into what has become known as *Population III* (or *Pop III*). As often happens in astronomy, this is a case of counting 'backwards': Population III is the term used for the primordial generation of stars in the universe.

These Pop III stars are in principle characterized by a near-absence of

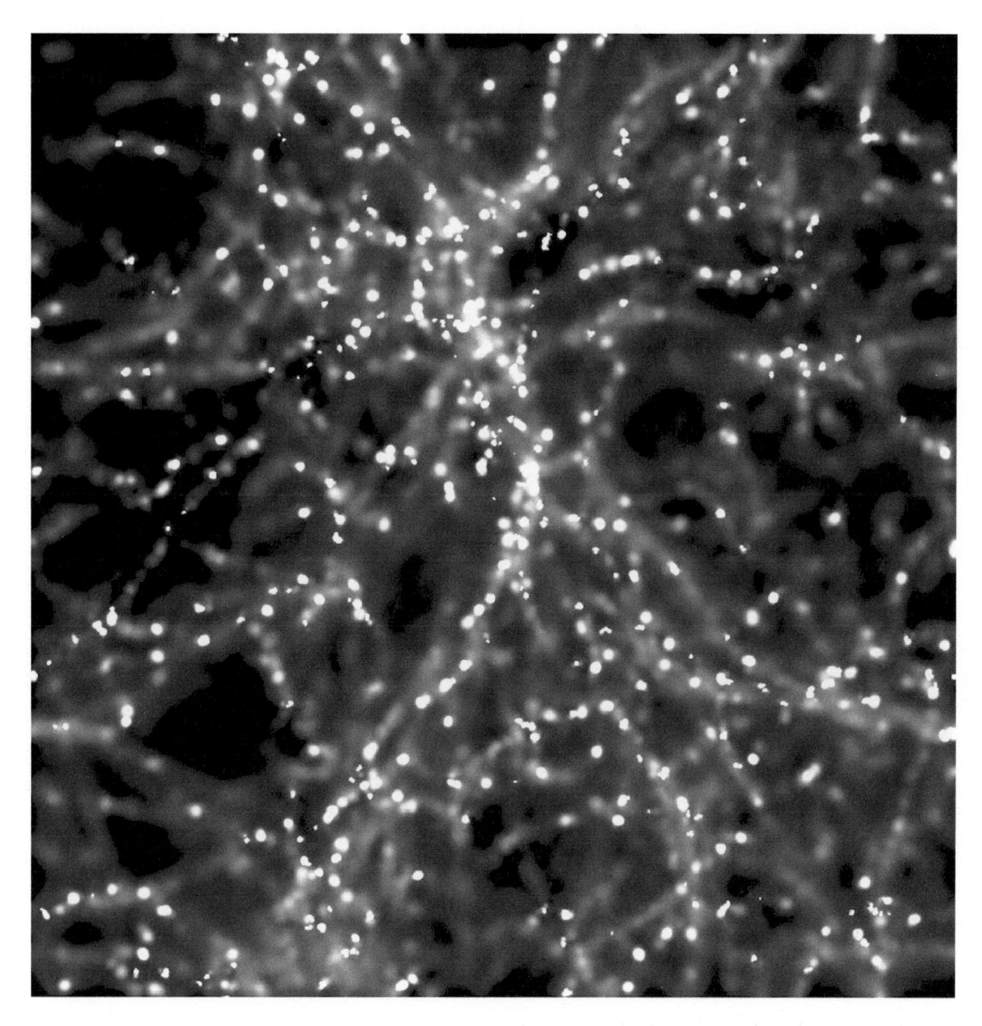

Figure 7.4 Spatial distribution of neutral hydrogen at redshift $z \sim 2$, obtained digitally in a simulation representing a cubic volume of the universe about 10 million light years across. Note the presence of many isolated clouds, which will be responsible for the absorptions causing the Lyman 'forest'.

metals,[2] since they were created from the products of primordial nucleosynthesis: hydrogen and helium.

These very massive stars (up to several hundred times the mass of the Sun) came to a rapid end and were the first to enrich the interstellar medium, having

2. In astrophysical terms, 'metals' are all the elements heavier than hydrogen (referred to as *X*) and helium (*Y*). Metallicity (*Z*) represents the proportion (in number and in mass) of 'heavy' atoms.

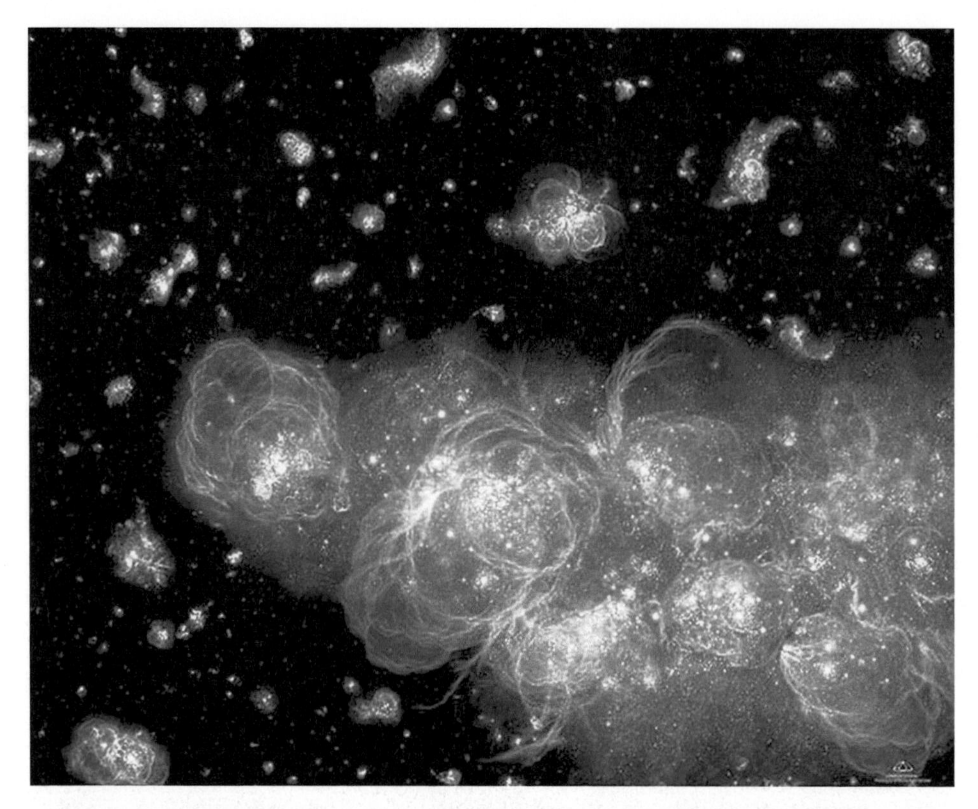

Figure 7.5 This is an artist's impression of how the very early universe (less than 1 billion years old) might have looked when it went through a voracious onset of star formation, converting primordial hydrogen into myriad stars at an unprecedented rate. The most massive of these Population III stars self-detonated as supernovae, which exploded across the sky like a string of firecrackers. (Adolf Schaller for STScI.) See also PLATE 21 in the color section.

synthesized all the elements up to iron (Figure 7.5). Population II stars, also of low metallicity, were formed from their 'ashes'. Over millions and billions of years, the cycle went on, until the stars of Population I, one of which is our Sun, shone as the heirs to their ancestors' store of metals. Metallicity is therefore used to estimate the age of any given star.

So where did Pop III go?

To date, no unequivocal candidate for a Pop III star has been discovered. Their existence remains pure conjecture, but, as we have seen, there must have been a first generation, for successive generations to have existed.

One of the probable reasons for this blank page in the cosmic census is that here we are dealing with stars which are not only very ancient but also had very short lifetimes, of less than a million years. No trace remains of them in today's

Figure 7.6 An artist's impression of the James Webb Space Telescope (JWST), a large infrared telescope with a 6.5-meter primary mirror. Launch is planned for 2013. JWST will be the premier observatory of the next decade, serving thousands of astronomers worldwide. It will study every phase in the history of our Universe, from the first luminous glows after the Big Bang, to the formation of solar systems capable of supporting life on planets like Earth. JWST was formerly known as the 'Next Generation Space Telescope' (NGST), but it was renamed in September 2002 after a former NASA administrator, James Webb. (NASA/ESA.) See also PLATE 22 in the color section.

universe, and if we are to look for them, it must be at very great redshifts. Two possible methods: either extremely sensitive instruments will have to be deployed to carry out this search (for example the James Webb Space Telescope (Figure 7.6), a successor to the Hubble), or we must observe objects such as GRBs during the next decade, out to redshifts of 10–15 (i.e. to a time when our universe was less than 500 million years old), as they seem to be serious Pop III candidates. No doubt both methods will be tried.

The outstanding advantage will be that the first stars can be observed, one by one.

A family group

Understanding in detail and quantifying the chronology of the episodes of star formation, i.e. identifying Populations I, II and III and their relationships, is one of the key questions in cosmology and in the study of the scenario of the formation of cosmic structures. Great progress has been made in this field in recent years, thanks to considerable improvements in instruments working in the infrared and sub-millimetric domains: observations (both ground-based and from space) have in fact revealed the existence of a large population of objects invisible in the optical region, as they are 'buried' in dust and therefore all but lost to view.

GRBs, with their intense gamma-ray bursts followed by a long period of emissions (from X-rays to radio waves) which decline gradually, are hardly affected by this dust. They are observable at great distances and offer clues to the 'tragic ends' of massive stars, so studying them, and their host galaxies, at various redshifts (i.e. at various times in cosmic history) provides us with information about one of the universe's three star-families.

From recombination to reionization

A crucial episode in the thermal history of the cosmos (see Chapter 2) is the period known as the 'Recombination Era', during which, in spite of its name, protons and electrons combined for the first time to form atoms. As we have seen, this neutralization of the cosmic plasma allowed photons, hitherto trapped in their interactions with electrons, to escape as they finally diffused from the charged particles. They were then free to travel through the universe, forming the *cosmic background radiation* observed by COBE and, more recently and in greater detail, by WMAP. The extra information from WMAP involved the fluctuations in the temperature of this radiation: the universe, it was revealed, had been *re-ionized* some time after the period of recombination[3] (see summary in Figure 7.7).

The confirmation (or otherwise) of this episode and its details has become a crucial test for the cosmological model and could attain the status of a new 'pillar' of the Big Bang.

Indeed, observations of the spectra of quasars had already suggested that such an episode of re-ionization might have occurred. In the mid-1960s, astronomers James Gunn and Bruce Peterson conceived a test of the reality of this phenomenon. Just after the recombination, the universe was essentially a homogenous collection of hydrogen atoms, absorbing all the ultraviolet photons which might have been present and causing this period of the history of the universe to be totally opaque to our telescopes (hence its label of 'the Dark Ages'). After a phase of expansion and cooling, the processes leading to the formation of cosmic structures gradually asserted themselves, and matter condensed into the

3. The dating of this episode remains difficult, and proposed redshifts vary from 6 to 15, i.e. 900 to 300 million years after the formation of the universe.

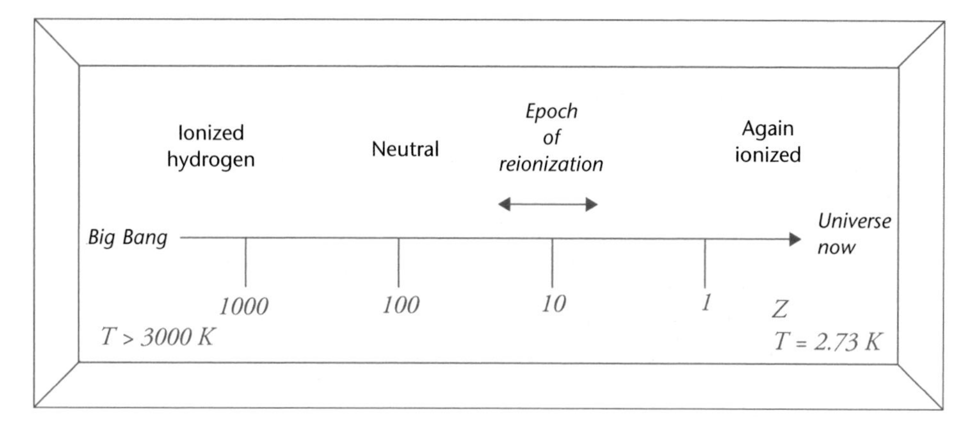

Figure 7.7 The history of cosmic plasma before and after recombination, indicating a phase of reionization some time after the recombination and the emission of cosmic fossil radiation. Determining the moment when the reionization period begins has become an essential test of cosmological models.

first luminous objects (as we saw in Chapter 3). It may have been that the radiation emitted by these primordial objects was energetic enough to locally (re-) dissociate atoms into protons and neutrons.[4]

However, the ionized medium having been created, became greatly diluted due to the cosmic expansion becoming transparent to photons. It was very inhomogeneous, because it consisted of ionized 'bubbles' around the objects formed. It remained so, as all the bubbles did not join up, leaving scattered 'residual pockets' of neutral gas. It occurred to Gunn and Peterson that if, in every 100 000 atoms dissociated by the radiation, even one neutral atom remained, the absorptive ability of the neutral atoms would noticeably alter the spectrum of a quasar. This spectrum, to a first approximation consisting of 'continuous radiation' upon which are superposed lines such as the Lyman-α, as already mentioned, is affected by everything lying along the line of sight.

Without an absorbing medium, the 'ultraviolet continuum of the spectrum' (i.e. the part further towards the blue than the Lyman-α), remains unaffected, and is therefore similar to the continuum beyond the Lyman-α. The appearance of a trough in the blue part of the spectrum indicates that the medium traversed is neutral and not ionized.

Evidence for such an effect was finally obtained in the twenty-first century, by an SDSS team studying quasars at redshifts greater than 6: quasars closer to us exhibit a continuum which is similar on either side of the Lyman-α. This indicates an ionized medium.

---

4. The so-called 'Strömgren spheres', dilating around a star such that the radiation emitted can ionize the surrounding medium.

Our current universe is indeed, therefore, in an ionized phase, and observations of quasars at greater and greater distances from us, which may or may not show the trough, allow us in theory to establish the epoch when this re-ionization began.

Comparisons of the results obtained by this technique, using measurements from WMAP, seem to indicate a good agreement, though there is as yet no definitive confirmation.

To achieve the most accurate possible dating, we have to observe even more distant and more luminous objects: gamma-ray bursts once again come to the rescue, and measuring their spectra with a new generation of instruments will allow us to obtain key signatures.

8 A bright, dark future

'Time steals away all things, even the mind'

Virgil, classical Roman poet

The oxymoron in the title of this chapter is in itself a good summary of the content of this book.

The future will indeed be bright, as coming decades see the introduction of very large instruments, both ground-based and in space, linked to very ambitious observational projects which will take us further towards the goals of precise cosmology.

Bright, too, because such studies will make use of the most luminous objects that the universe has produced in the whole of its history.

On the theoretical side, the future also looks bright, for it may well be that the consequences for fundamental physics will be revolutionary, marrying – it is to be hoped – general relativity to quantum physics.

And the dark side of the future? Unless there is a sea-change in our appreciation of the universe, it is and will remain dark in more than one sense of the word. Dark with the darkness of the matter governing the dynamics of galaxies and clusters of galaxies; and dark with the darkness of the energy dominating the expansion of the cosmos.

We shall now embark upon the risky but necessary exercise of trying to look into that future, and sketching out a panorama for the decades – and even the millennia – to come.

A bright future for observers

The (long-term) hunt for supernovae

In order to compare the various theoretical hypotheses, we need continuously to improve the quality of our observations, and even pursue new observational directions. As far as the search for Type Ia supernovae is concerned, there are several possible paths to be explored.

First and foremost, we must maintain our efforts using current methods of observation, but in new ways. One exciting aspect is research into very-high-redshift supernovae, looking back to an epoch when our universe was still very young. Such observations are critical, because we can use them, for example, to estimate the influence of dust in our interpretation of results: since the distance

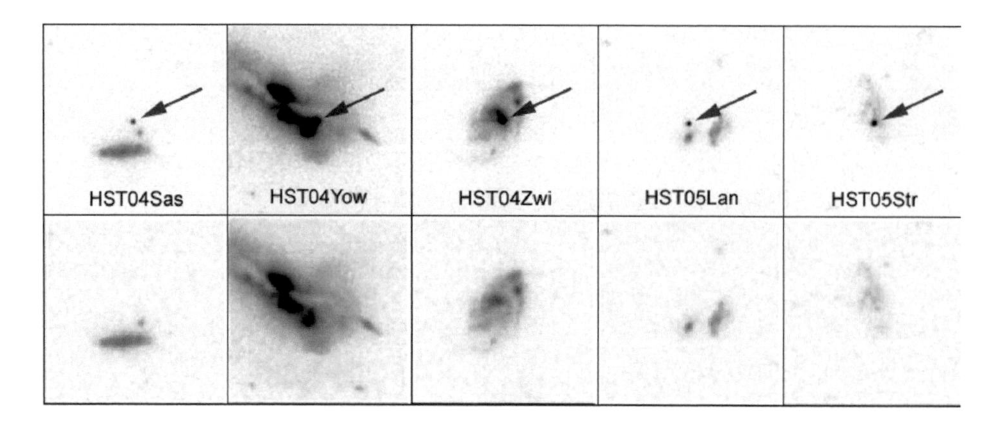

Figure 8.1 Images of Type Ia supernovae discovered and monitored using the Hubble Space Telescope. The supernovae (arrowed) with their host galaxies are shown in the upper images, with reference images showing only the host galaxies below them. Images taken by instruments in space, and therefore not subject to problems caused by the Earth's atmosphere, allow us to detect and monitor very faint (and therefore very distant) supernovae. (NASA.)

between us and the objects observed is very great, the contribution of dust is that much more important. For the same reason, it is imperative that we continue to compare the properties of very distant supernovae with those nearer to us, in order to ascertain whether these 'candles' are indeed 'standardizable' and whether they change with time. These observations will mean that we can at last explore the transitional period between the phases of deceleration and acceleration of the expansion, giving us a clearer picture of cosmological scenarios.

The Hubble Space Telescope (HST), for example, has made it possible to study supernovae that exploded between 3 and 10 billion years after the formation of our universe (Figure 8.1). At present, it is the one instrument in space, far above the hindrances caused by the Earth's atmosphere, capable of making such observations: the objects in question are much too faint to be detectable by even the largest Earth-based telescopes. However, unlike (for example) the MegaCam imager of the Canada-France-Hawaii Telescope (CFHT), the HST's instruments have a field of view that is much too small[1] for systematic research programs to be carried out with them.

Another very promising field of modern research involves the development of studies of supernovae in the infrared; the limited number of observations to date show that, in this wavelength domain, Type Ia supernovae could serve as

1. For example, the Advanced Camera for Surveys (ACS), the largest instrument ever installed on the Hubble Space Telescope, covers an area of sky more than 300 times smaller than that available to MegaCam/CFHT.

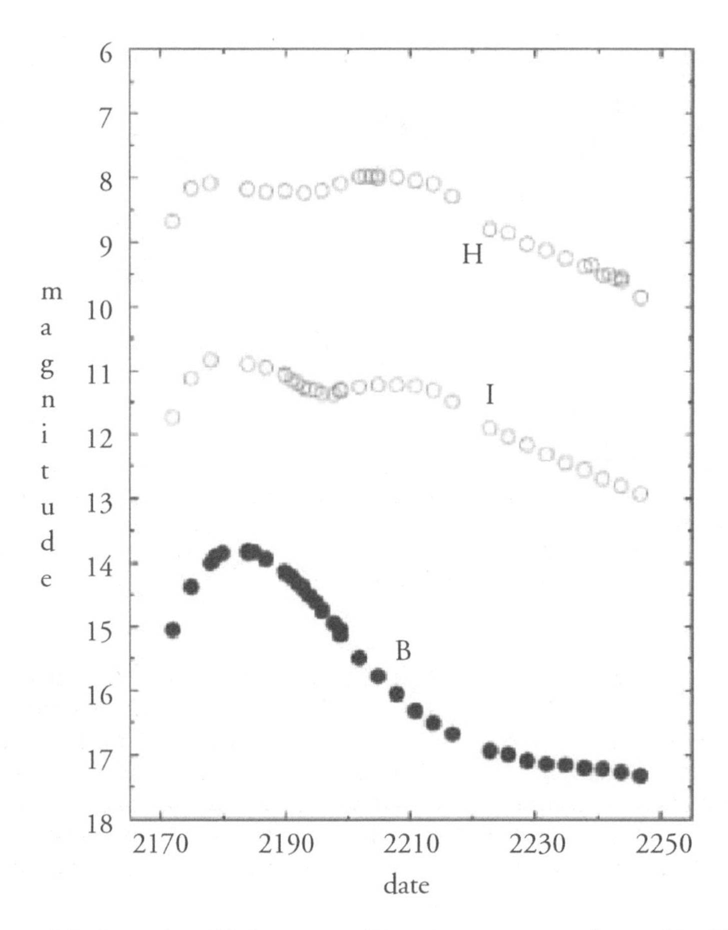

Figure 8.2 Examples of light curves of Type Ia supernovae observed in different spectral domains. The evolution of the luminosity over time is visible in the blue (B), the red (I) and the infra-red (H). Note that the shape changes as a function of the spectral domain. In particular, the fact that the 'redder' (I and H) curves are 'flatter' means that we do not have to determine the date of maximum very precisely, and that a better standardization of the supernovae could be possible.

much better 'standard candles' than in the visible part of the spectrum. The maxima of their light curves are in fact plateaux (Figure 8.2), which means that it becomes easier to determine their precise maxima. Further infrared monitoring programs will be undertaken in the years to come, using wide-field imagers on large telescopes; new insights should be gained in this field.

Another and equally obvious avenue is that of the development of new observational methods. This is the classic situation whereby new scientific discoveries often hasten the need for new instruments.

In the context of the hunt for supernovae, it is tempting to exploit the

Figure 8.3 Artist's impression of the SNAP (SuperNova Acceleration Probe) satellite, which will be able to detect and observe several thousand supernovae out to redshifts of approximately $z \sim 1.7$. It will also be able to measure gravitational lensing effects created by the distribution of dark matter in the universe. Such measurements will be used to trace out the history of cosmic expansion over the last 10 billion years, and determine, in conjunction with measurements of the cosmic background radiation by the European Planck satellite, the nature of dark energy and its possible evolution over cosmic time. (SNAP/LBL.) See also PLATE 23 in the color section.

qualities of observations made by instruments in space[2] in systematic research programs into Type Ia supernovae.

One example of this is the proposed SNAP satellite (Figure 8.3); the aim of this

---

2. The absence of the atmosphere ensures: 1) an image quality ($\sim$0.1 seconds of arc) much better than that obtained from ground-based instruments ($\sim$0.9 seconds of arc); 2) a drastic ($\sim$1/20) diminution of the 'sky background'; and 3) freedom from atmospheric effects.

project is to sample several thousand supernovae out to a redshift of 1.7 (corresponding to an age of the universe of 4 billion years), and to couple these observations to other cosmological probes such as the 'gravitational shear'. Images taken from space, at regular intervals, of the same area of the sky will ensure coverage of, and detection in, a very wide spectral range including the visible and the near-infrared.

Certainly, definitive progress will be made by a project such as SNAP, or its equivalent, in conjunction with the European Planck satellite, expected to obtain improved measurements of cosmological radiation, or the James Webb Space Telescope (successor to the HST). Unfortunately, space telescope projects are always technically challenging and very costly. It takes a long time to bring them to fruition, and a project such as SNAP will probably not see the light of day until 2015. Science and patience always proceed hand in hand.

Desperately seeking a burst...

The Earth's atmosphere efficiently absorbs gamma-rays, protecting life on this planet from their harmful effects. However, it also prevents ground-based astronomers from discovering gamma-ray bursts. To achieve this end, they are obliged to launch satellites, observing from above the atmosphere.

The most important such mission now in progress is undoubtedly the Swift satellite, launched in November 2004 by NASA. Since gamma-ray bursts can occur in any part of the sky, a very wide area[3] must of course be covered, by a telescope working in the gamma-ray zone of the spectrum.

A specific feature of this satellite is that, once it has detected a burst, it automatically points towards it, taking just a few minutes to do so; then its X-ray telescope begins to operate, 'homing in' on the position of the object on the celestial sphere (remember that it is difficult to pinpoint the exact origin of gamma-rays). This technical feat means that it is possible to locate the position of a burst with remarkable accuracy, to within a few hundredths of a degree. Swift therefore takes forward the legacy of the successful observations (1996-2002) of the Italian-Dutch satellite BeppoSAX.

Swift has worked with remarkable efficiency ever since its launch. It detects and observes more than 90 bursts every year, and thanks to Swift, we have proof of the considerable distances at which gamma-ray bursts can occur: currently the record stands at a redshift of 6.7, that is to a time when our universe was less than one billion years old, as we have seen in Chapter 5.[4] Another particularly notable result involves our understanding of short bursts. It seems very likely that these

3. Typically more than 5 000 square degrees (about one-eighth) of the sky, if at least one burst per week is to be detected.
4. The current record holder is GRB 080913, detected on 13 September 2008, during the final reading of the English language edition of this book. When this event occurred, the universe was only about 800 million years old.

are linked to the coalescence of compact objects such as black holes or neutron stars. Unless, as can never be ruled out with this type of project, some major fault develops with the satellite, Swift should continue to function for many years until some worthy successor comes along.

Just as in the case of the hunt for supernovae, it is necessary to bring to bear existing instruments at the same time as we prepare new projects which encompass the latest advances. The most remarkable progress made in recent years in this field has resulted from very close collaboration between instruments in space (to detect and pinpoint bursts) and ground-based detectors (measuring redshifts and identifying host galaxies); with this in mind, France and China are pursuing an ambitious satellite project, its highly promising axes of research being the understanding of the phenomenon and its use in cosmology.

A combination of space- and ground-based instruments on a hitherto unrivalled scale will mean that this satellite should be able to study gamma-ray bursts across a very wide spectral range, from gamma-rays through visible light to the near infrared; in theory, it will be able to observe the very first stars that formed in our universe: the famous Population III, looking back to a time when those stars were less than 500 million years old (a redshift of the order of $z \sim 10$). Patience is still the watchword: the satellite in question will probably not be launched before 2013.

And more ambitious still...

Researchers are always coming up with new ideas, and we could go on to mention many other projects; however, this is not meant to be *The Book of Lists*, so it might be of more interest to finish our brief *tour d'horizon* by looking at one of the highest priorities in world astronomy: the need for very large telescopes.

Now, in the quest to observe ever more remote (and ever fainter) objects, we need to collect more and more light. Today's largest telescopes are from 8 to 10 meters in diameter. In the near future, a new era of much larger instruments will be upon us, with telescopes more than 30 meters in diameter. This represents a 'normal' step along the observational road: Figure 8.4 shows how the diameter of ground-based telescopes has increased exponentially as techniques have improved.

There are several such projects in progress at present. These include the Thirty Meter Telescope (TMT), a joint effort by the USA and Canada (Figure 8.5), and the European Extremely Large Telescope (E-ELT), up to 42 meters in diameter, which is undoubtedly very ambitious (Figure 8.6). Due to see first light at the end of the next decade, the E-ELT will be of particular value in observations of supernovae associated with the earliest (Pop III) stars.

The next two decades will therefore see the advent of very large instruments, both on Earth and in space. A huge harvest of data will come from them, opening the way to a veritable 'precision cosmology' and an ever more refined picture of the properties of our universe.

So, a bright future is in store for the observers. But... will we be able to interpret this flood of data correctly? This is the question which will accompany the theoretical developments of the future.

PLATE 1 The expanding shell of debris from the supernova which was seen to explode in 1006. This image is a composite of visible (or optical), radio, and X-ray data of the full shell of the supernova remnant. (NASA, ESA, and Z. Levay (STScI).)

PLATE 2 This composite image of the Crab Nebula uses data from the Chandra X-ray Observatory, Hubble Space Telescope, and the Spitzer Space Telescope. The central neutron star – the remains of the star which was seen to explode in 1054 – is the bright white dot at the center of the image. (NASA, ESA, CXC, JPL-Caltech, J. Hester and A. Loll (Arizona State Univ.), R. Gehrz (Univ. Minn.), and STScI.)

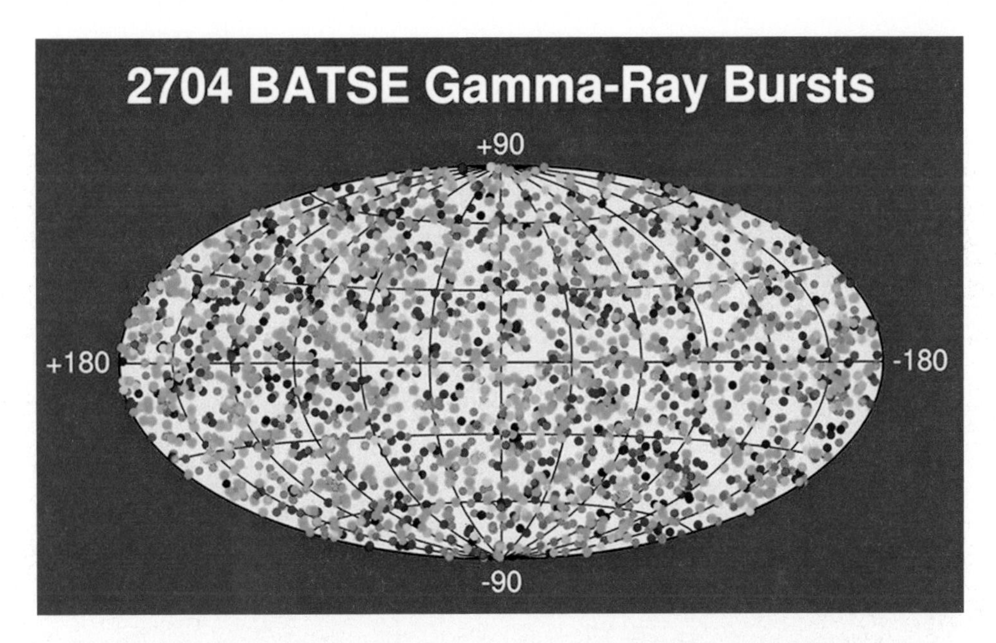

PLATE 3 Chart showing the positions of 2704 bursts observed by the BATSE instrument on board the Compton Gamma-Ray Observatory. The whole sky is shown, in galactic coordinates (the galactic centre being at 0°/ 0°). (NASA Marshall Space Flight Center.)

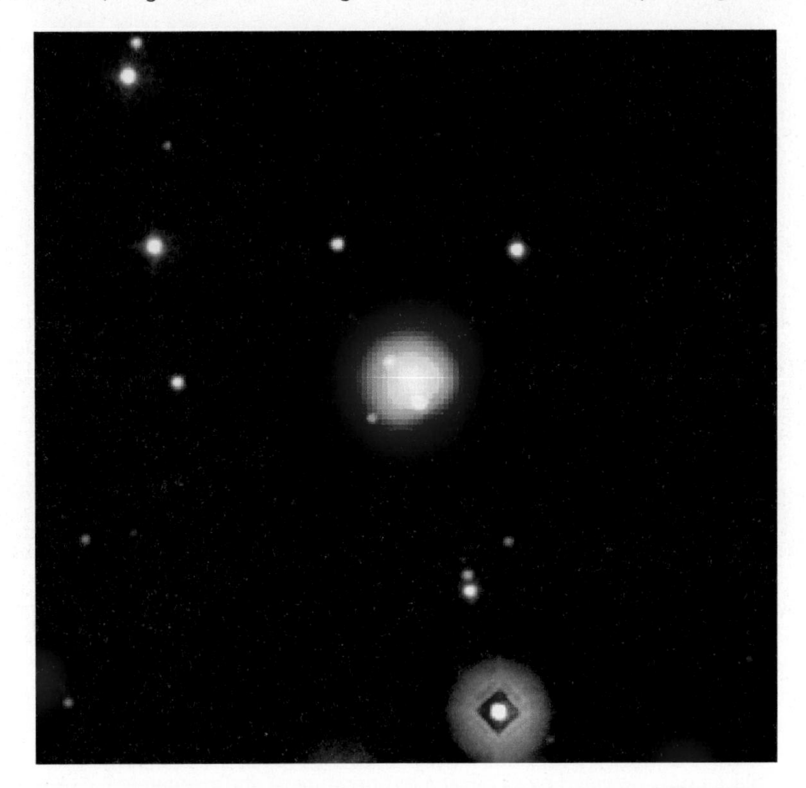

PLATE 4 This image of GRB 080913, the most distant gamma-ray burst recorded to date, merges the view through Swift's UltraViolet and Optical Telescope, which shows bright stars, and its X-ray Telescope, which captures the burst, visible near the center of the image. (NASA/Swift/Stefan Immler.)

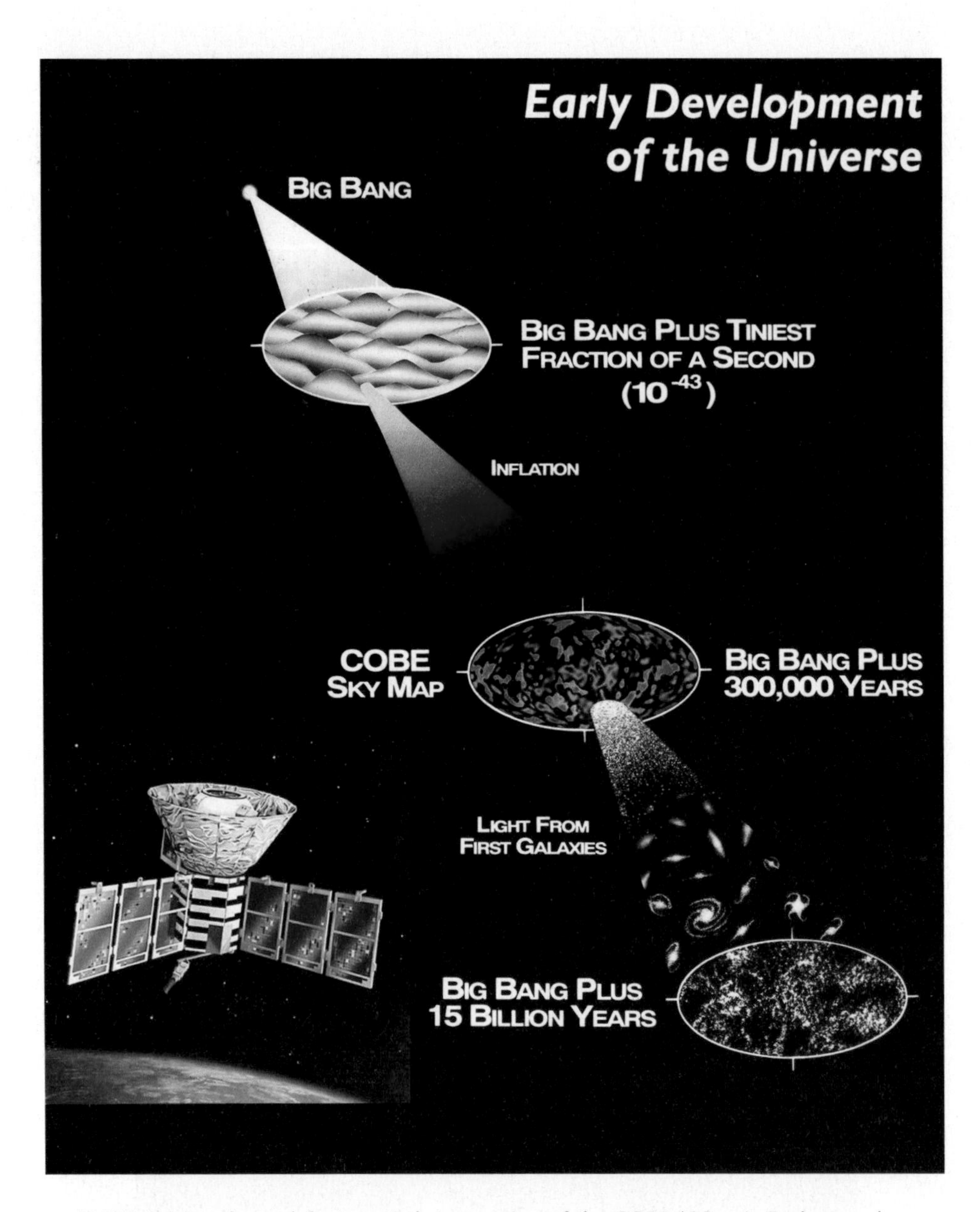

PLATE 5 (a) – lower left: An artist's impression of the COBE (COsmic Background Explorer) satellite in Earth orbit. **(b)** – right: Cosmic history since the Big Bang. After the Planck time (about 10^{-43} s), there occurred an inflationary phase during which the universe expanded exponentially. The primordial fluctuations which gave rise to the galaxies we see today were generated at this epoch. After 380,000 years, photons decoupled from matter and flooded out freely through the universe. They constituted a perfect black body, detected by COBE at a temperature of approximately 2.73 K. (NASA Goddard Space Flight Center.)

PLATE 6 This million-second-long exposure, called the Hubble Ultra Deep Field (HUDF), reveals the first galaxies to emerge from the so-called 'Dark Ages', the time shortly after the Big Bang when the first stars reheated the cold, dark universe. This view is actually two separate images taken by Hubble's Advanced Camera for Surveys (ACS) and the Near Infrared Camera and Multi-object Spectrometer (NICMOS). (NASA, ESA, S. Beckwith (STScI) and the HUDF Team.)

PLATE 7 NASA's Hubble Space Telescope captures the magnificent Coma Cluster of galaxies, one of the densest known galaxy collections in the universe. Hubble's Advanced Camera for Surveys viewed a large portion of the cluster, spanning several million light-years across. (NASA, ESA, and the Hubble Heritage Team (STScI/AURA).)

PLATE 8 This Hubble Space Telescope image of a rich cluster of galaxies called Abell 2218 is a spectacular example of gravitational lensing. This cluster of galaxies is so massive and compact that light rays passing through it are deflected by its enormous gravitational field. This phenomenon magnifies, brightens, and distorts images of more distant objects. (Andrew Fruchter (STScI) et al., WFPC2, HST, NASA.)

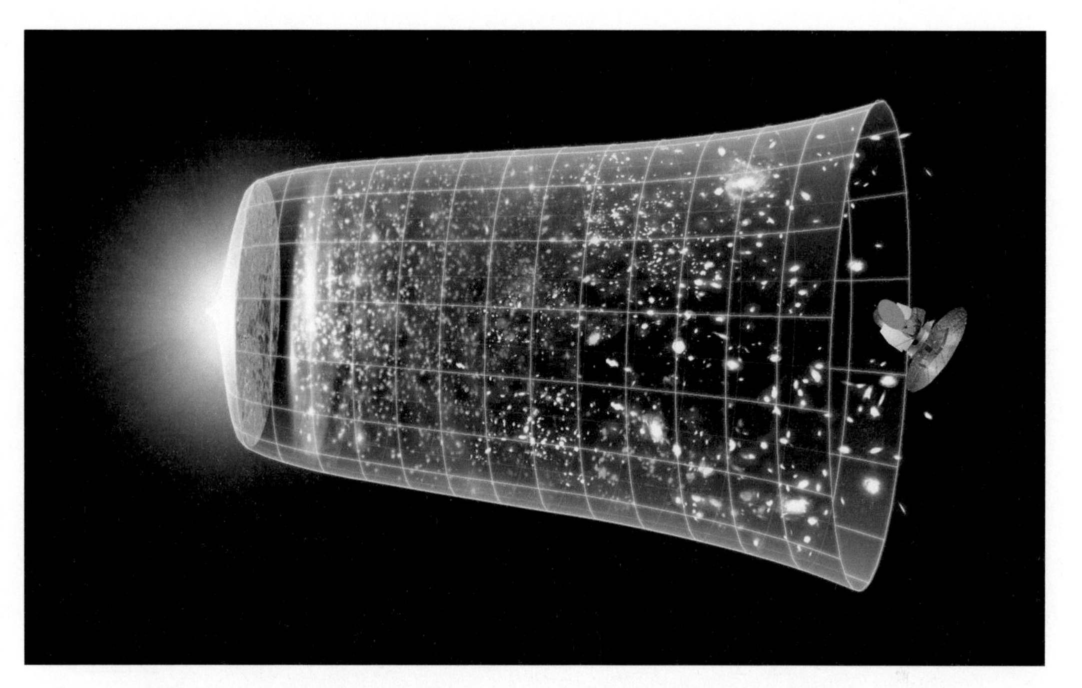

PLATE 9 A representation of the evolution of the universe over 13.7 billion years. The far left depicts the earliest moment we can now probe, when a period of 'inflation' produced a burst of exponential growth in the universe. The afterglow light seen by WMAP was emitted about 380,000 years after inflation. (NASA/WMAP Science Team.)

PLATE 10 The detailed, all-sky picture of the infant universe from three years of WMAP data. The image reveals 13.7 billion year old temperature fluctuations (shown as differences in tint) that correspond to the seeds that grew to become the galaxies. This image shows a temperature range of $\pm$ 200 microKelvin. (NASA/WMAP Science Team.)

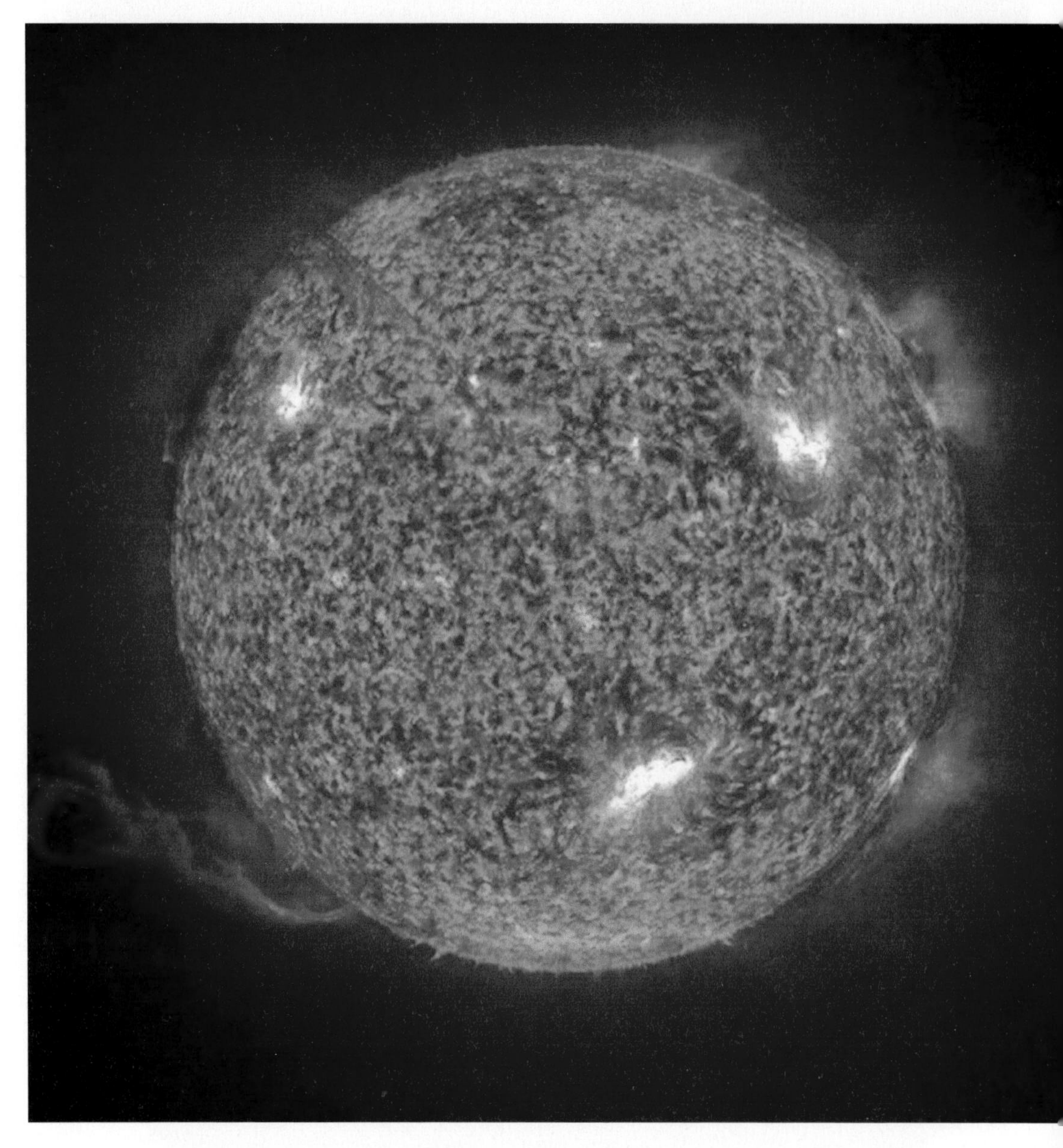

PLATE 11 This X-ray image of the Sun, taken by the SOHO satellite, shows numerous active regions in the Sun's atmosphere. The hottest and most active regions appear white, and the darker areas indicate cooler temperatures. The wispy feature in the lower left portion of the disk is a solar prominence, a huge cloud of relatively cool plasma suspended in the Sun's hot thin corona. (SOHO (ESA + NASA).)

PLATE 12 Expanding light echoes illuminate the dusty surroundings of V838 Monocerotis, a mysterious red supergiant star near the edge of our Galaxy. This image was produced from Hubble Space Telescope data recorded in October 2004. (NASA, ESA, The Hubble Heritage Team (AURA/STScI), and H.E. Bond (STScI).)

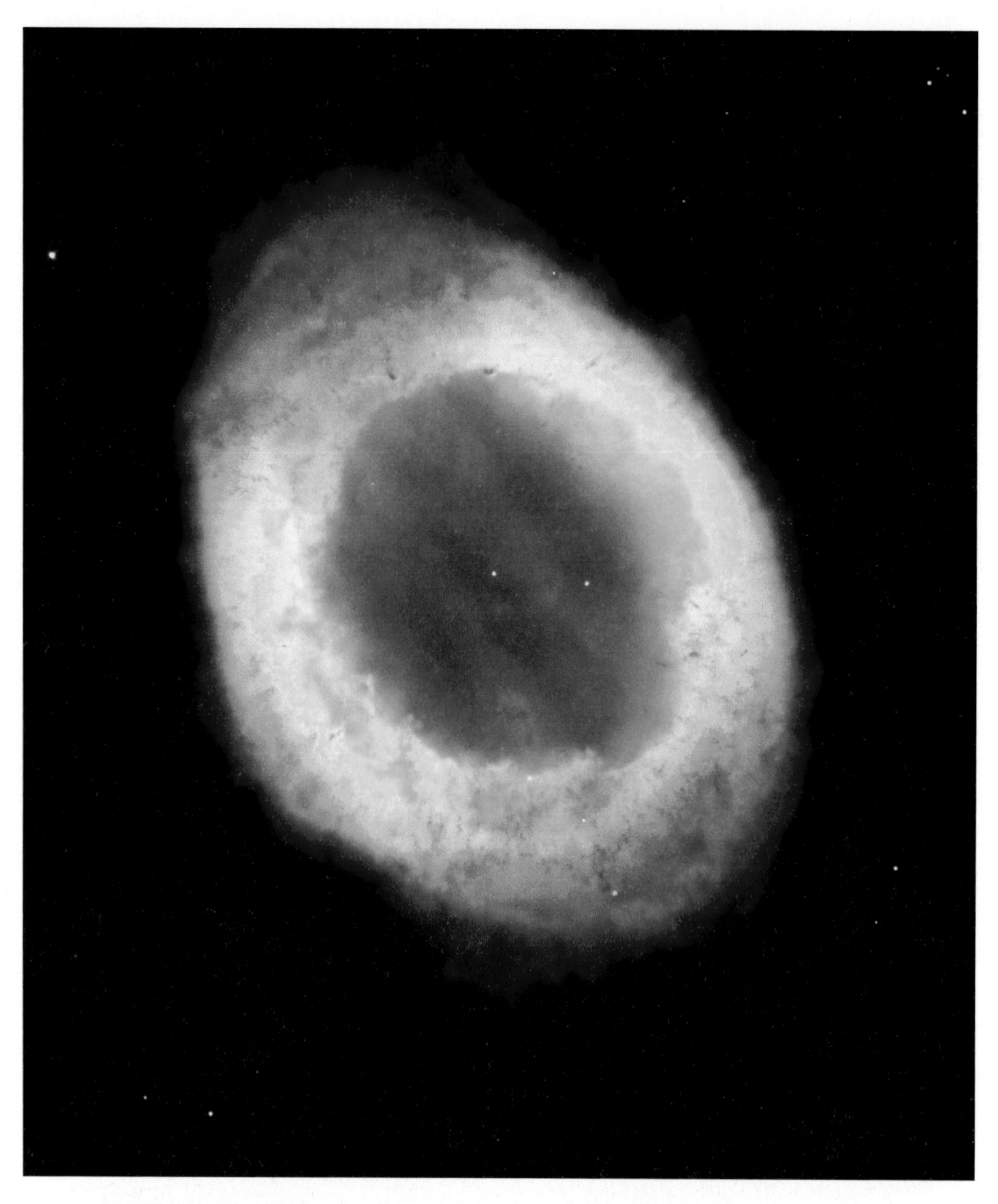

PLATE 13 A Hubble Space telescope image of the Ring Nebula. The faint speck at its center was once a star of greater mass than our own Sun. Now, near the end of its life, it has ejected its outer layers into space, and the remnant is a tiny white dwarf star, about the size of the Earth. (The Hubble Heritage Team (AURA/STScI/NASA).)

PLATE 14 The enclosures of the four unit telescopes of the Very Large Telescope (VLT) at the European Southern Observatory (ESO) on the summit of Cerro Paranal in Chile.

PLATE 15 The Fireworks Galaxy NGC 6946. This is a relatively nearby, face-on spiral galaxy to our Milky Way, located just 10 million light-years distant. Looking from the bright core outward along the loose, fragmented spiral arms, there is a change from the light of old stars in the galaxy's center to young star clusters and star forming regions. NGC 6946 is rich in gas and dust, with a high star birth and death rate. In the past century, at least eight supernovae were discovered in this galaxy. Their positions are marked here.

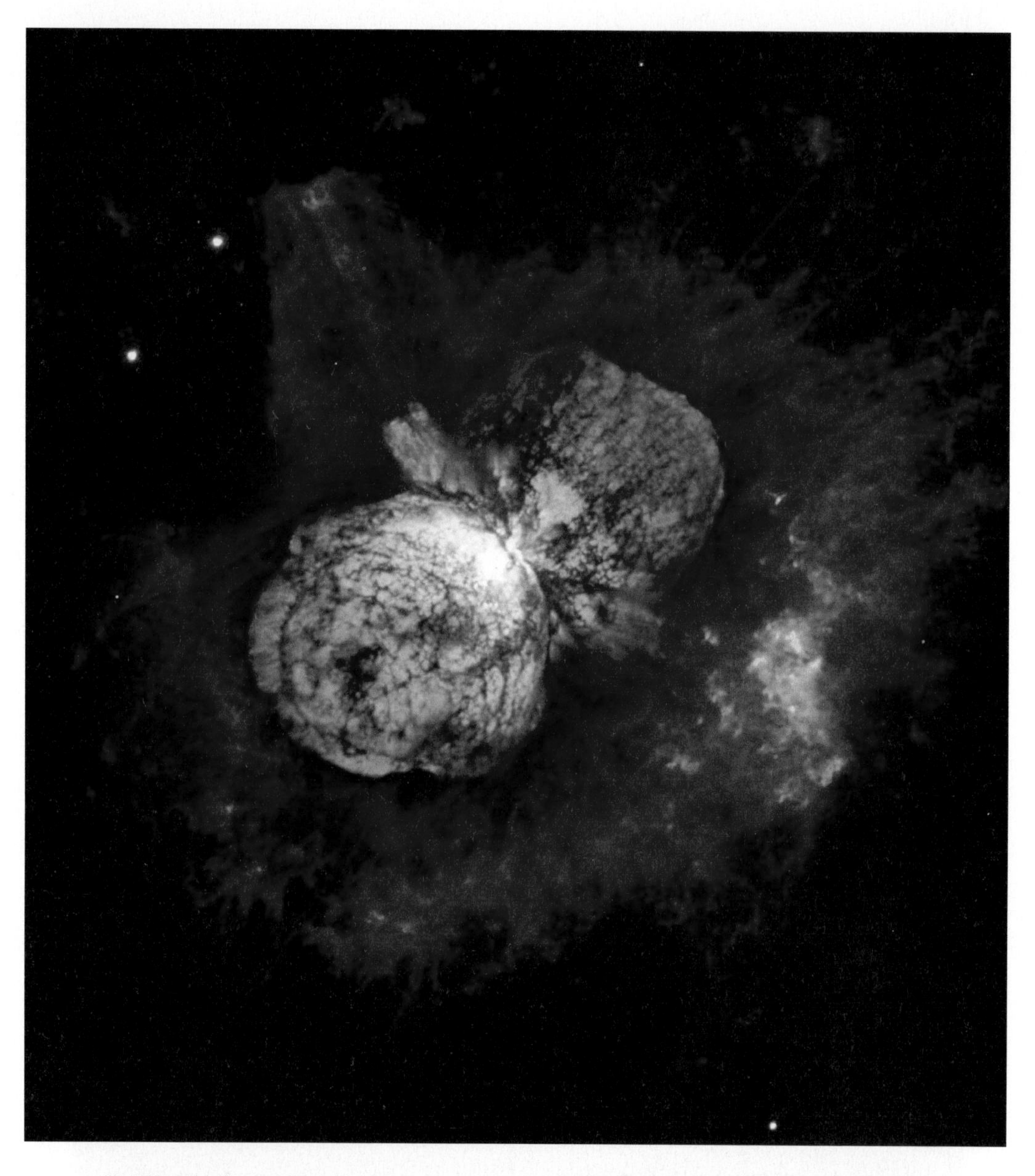

PLATE 16 About 170 years ago, the southern star Eta Carinae mysteriously became the second brightest star in the night sky. In 20 years, after ejecting more mass than our Sun, Eta Carinae unexpectedly faded. This outburst appears to have created the Homunculus Nebula, pictured here in a composite image from the Hubble Space Telescope. Eta Carinae still undergoes unexpected outbursts, and its high mass and volatility make it a candidate to explode in a spectacular supernova sometime in the next few million years. (N. Smith, J. A. Morse (U. Colorado) et al., NASA.)

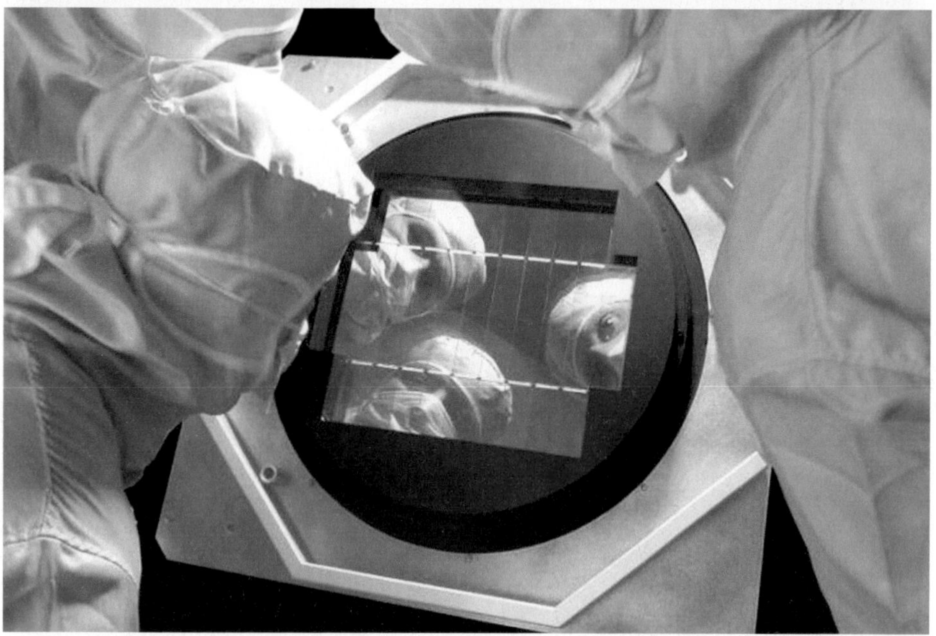

PLATE 17 (a) – top: To observe distant supernovae, astronomers need to use telescopes of at least 4 meters aperture, such as the CFHT (Canada-France-Hawaii Telescope) in Hawaii, shown here at night with star trails behind, both to discover them and to monitor their light curves. (Jean-Charles Cuillandre (CFHT).) **(b)** – bottom: Associated with these telescopes are wide-field cameras, such as MegaCam, which covers about 1 square degree of sky, and is at present one of the largest imagers in the world. (Paris Supernova Cosmology Group.)

PLATE 18 Edwin Hubble's name was given to the space telescope, eventually launched in April 1990, whose rich harvest would revolutionize astrophysics in all its aspects. (NASA.)

PLATE 19 Identification and distance measurements of supernovae are carried out spectroscopically by the largest telescopes (8–10 meters) such as the twin Keck telescopes in Hawaii. (W.M. Keck Observatory.)

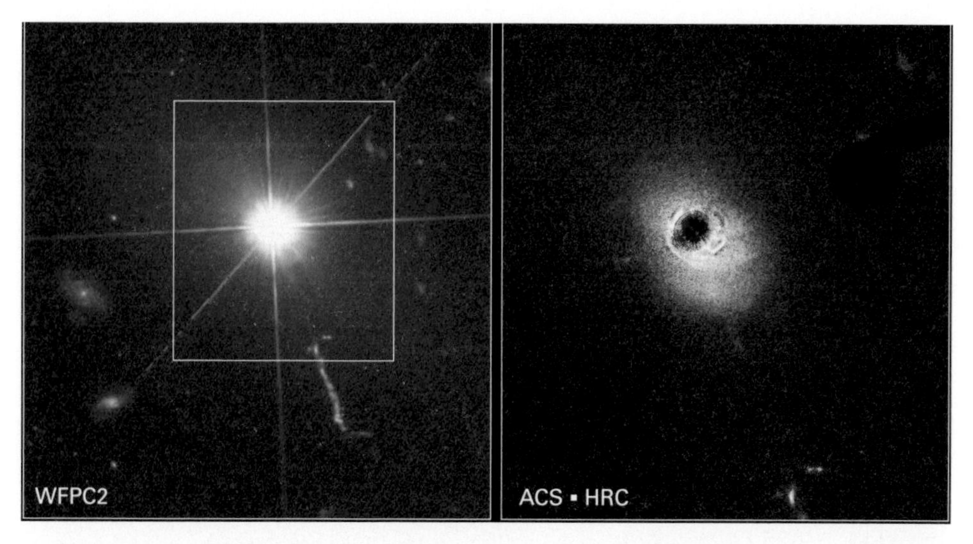

PLATE 20 Left: This image of the quasar 3C273 in the constellation of Virgo, acquired by the Hubble Space Telescope's Wide Field Camera 2, shows the brilliant quasar but little else. (NASA and J. Bahcall (IAS).) Right: Once the blinding light from the brilliant central quasar is blocked by the coronagraph of Hubble's Advanced Camera for Surveys (ACS), the host galaxy pops into view. (NASA, A. Martel (JHU), H. Ford (JHU), M. Clampin (STScI), G. Hartig (STScI), G. Illingworth (UCO/Lick Observatory), the ACS Science Team and ESA.)

PLATE 21 This is an artist's impression of how the very early universe (less than 1 billion years old) might have looked when it went through a voracious onset of star formation. The most massive of these Population III stars self-detonated as supernovae, which exploded across the sky like a string of firecrackers. (Adolf Schaller for STScI.)

PLATE 22 An artist's impression of the James Webb Space Telescope (JWST), a large infrared telescope with a 6.5-meter primary mirror. Launch is planned for 2013. It will study every phase in the history of our Universe, from the first luminous glows after the Big Bang, to the formation of solar systems capable of supporting life. (NASA/ESA.)

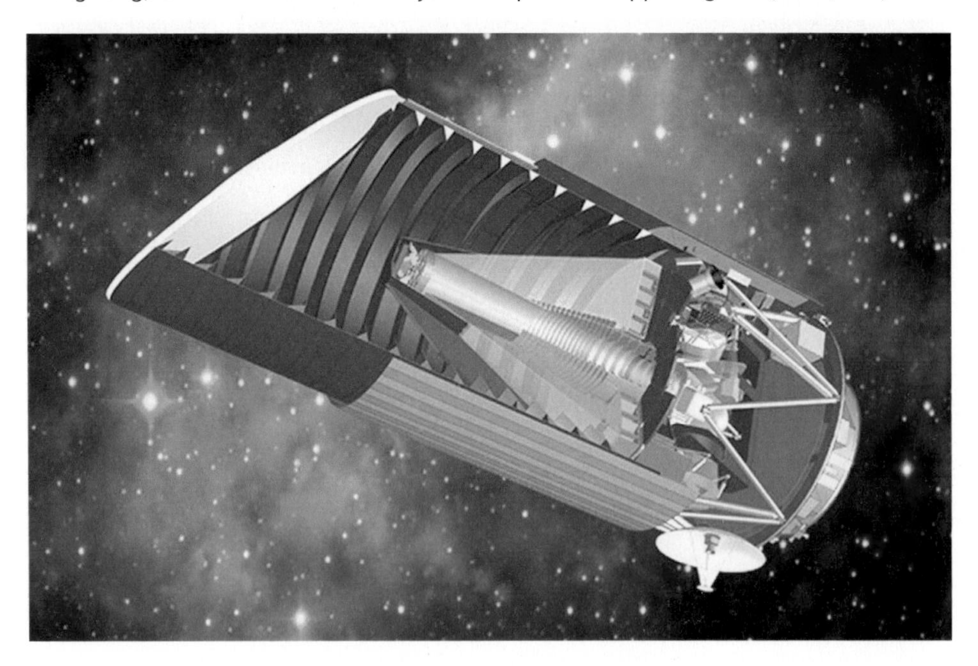

PLATE 23 Artist's impression of the SNAP (SuperNova Acceleration Probe) satellite, which will be able to detect and observe several thousand supernovae out to redshifts of approximately $z \sim 1.7$. It will also be able to measure gravitational lensing effects created by the distribution of dark matter in the universe. (SNAP/LBL.)

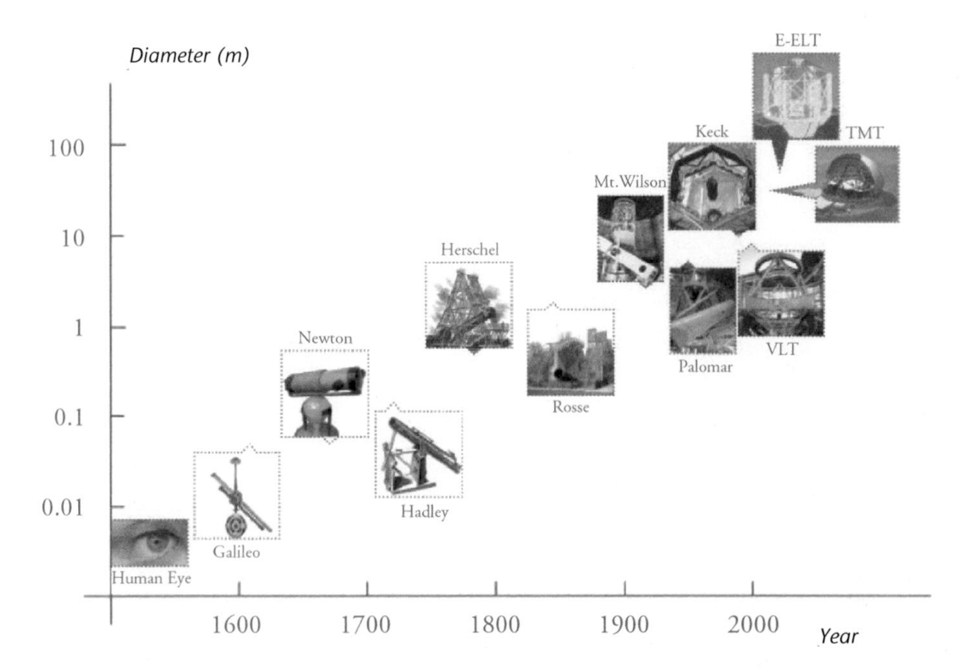

Figure 8.4 The historical increase in the diameters of ground-based telescopes. Note the exponential progression, as techniques improve. Every half-century, diameters increase by more than ten times! (ESO.)

A dark future for theorists

Cosmological constant or dark energy?

For the majority of cosmologists and physicists, the acceleration of the expansion of the universe is a well established phenomenon (though reservations are still expressed now and again). The most convincing argument probably lies in the well known cosmic triangle (mentioned earlier in this book). Measurements of cosmic 'black-body' radiation have established quite firmly that the total quantity of energy-matter Ω_{tot} is equal to unity (to within 2%, according to measurements by WMAP). Moreover, evaluations of the mass content (baryonic and dark) indicate a value of approximately 0.3 for the quantity of matter Ω_{mat}. It then becomes necessary to introduce a new component such that its contribution will be approximately 0.7, to satisfy the relationship between the three sides.

The simplest hypothesis, given our current knowledge, consists of identifying this component with the cosmological constant Λ, originally present in Einstein's equations.

Now, the presence of this new cosmic component in the equations governing the dynamic of the universe inevitably implies an acceleration of the expansion, independently of any standard candle or scale.

Figure 8.5 The Thirty Meter Telescope (TMT) is shown in this computer-generated illustration. The primary mirror consists of 492 separate hexagonal mirror segments, each controlled by individual actuators. The design expands on the successful operation of the 10-meter Keck telescopes. A centrally-located tertiary mirror directs light to the Nasmyth focus, located along the horizontal axis of the mount. (TMT.)

So we now possess a new cosmological paradigm within which the universe was first of all dominated by (mostly dark) matter, and then, from about 5 billion years ago, by dark energy or one of its manifestations. According to whichever of these components is dominant, there results, during the expansion, an interim phase of deceleration or acceleration.

Once the phase of acceleration is established, we naturally want to understand its origin, and this is of course where the real problems will arise. Two major hypotheses are currently in the frame: that of the cosmological constant initially postulated by Einstein, and the dark energy hypothesis possibly linked to quantum vacuum energy.

One way of distinguishing between the two is, as we have seen throughout

Figure 8.6 Artist's impression of the European Extremely Large Telescope (E-ELT) now being planned by the European Southern Observatory (ESO). With a 42-meter diameter primary mirror, the present baseline, its total rotating mass is 5500 tons. In order to appreciate the scale of this instrument, just look at the two tiny people at lower left! The two platforms on each side of the structure hold large instruments. The telescope features a novel, innovative design, based on 5 mirrors. The primary 42-meter mirror is composed of 906 segments, each 1.45 meters wide, while the secondary mirror is as large as 6 meters in diameter. A tertiary mirror, 4.2 meters in diameter, relays the light to the adaptive optics system, composed of two mirrors: a 2.5-meter mirror supported by 5000 or more actuators so as to be able to distort its own shape a thousand times per second, and one 2.7 meters in diameter that allows for the final image corrections. This five mirror approach results in an exceptional image quality, with no significant aberrations in the field of view. Given the technical difficulties and the cost of such a project, it will probably not be completed until the end of the next decade. (ESO.)

this book, to determine the equation of state characterizing these two types of cosmological 'fluids'.[5]

In practice it soon becomes apparent that using one single category of observational data will not establish this equation with sufficient accuracy. We therefore resort to 'combined analyses', with each analysis concentrating on a cosmological 'target' dependent in different ways upon the parameters we are seeking to evaluate.

Among these various 'targets' are: the well known Type Ia supernovae; gamma-ray bursts; fluctuations in the temperature of cosmological radiation observed by WMAP; the effects of weak gravitational lensing created by the distribution of dark matter; and the distribution of galaxies throughout the universe. We then assemble the different pieces of the puzzle (i.e. the independent measurements), taking into account their levels of uncertainty.[6] The intersection of our results gives the most probable value.

Even if there is as yet no definitive consensus, the results obtained by this kind of study are currently seen to be consistent with an equation of state corresponding to a cosmological constant.[7] If we believe this result to be definitive, the physicist may respond to it in two different ways.

The first response is to consider, as Einstein did initially, that we are dealing here with a 'cosmological constant' in the strict sense of the term, i.e. a constant term which fits naturally into the equations of General Relativity. We therefore interpret it as a new fundamental constant in physics linked with the 'geometrical part' of these equations. It would have equivalent status to other fundamental constants (for example, the gravitational constant G), and its value can be determined only through experiment, as is the case with the other constants; it could not be deduced from a fundamental theory. The second response consists in seeking out a physical nature other than that assigned to it by the 'geometrical' interpretation. We may consider that this term relates, not to the geometrical, but to the 'energy-matter' part of the same Einsteinian equations, and that it represents in this case a 'cosmological fluid' as well as matter and radiation. The properties of this fluid must certainly be very strange and run counter to all common sense: its density will remain constant in spite of the expansion, and its gravitational action is repulsive, being responsible for the acceleration of the expansion.

---

5. The equation of state of a fluid relates pressure p to density ρ by: $p = w\rho c^2$, where w may be a function of redshift z. The 'cosmological constant' or vacuum energy corresponds to the case $w = -1$. Other types of 'dark energy' correspond to $w \neq -1$, with w being possibly dependent on cosmic time.
6. Each measurement is affected by uncertainties which may be either statistical (for example, for a population N we have the error $\sqrt{N}$), or systematic (for example, the calibration of the flux from celestial bodies). These concepts are similar to those used in predictions of electoral pollsters.
7. i.e., in the most general equation of state $p = w\rho c^2$, the parameter w is equal to -1.

As paradoxical as this might seem, such a fluid is compatible with the laws of fundamental physics. In fact, the primordial universe around the Planck time must be considered as a system ordered by quantum mechanics. Quantum mechanics states that such a system must necessarily possess a non-zero fundamental level (the level of lowest energy), while having all the attributes of a cosmological constant: this is the *quantum vacuum*. Now, since the density of this 'vacuum' energy remains constant, unlike radiation and matter whose contributions decrease inexorably, it becomes possible to imagine that it might have regained dominance some billions of years ago, and is responsible for the observed acceleration.

Can we therefore conclude that the question is finally closed, and that the problem is solved? Alas, although the hypothesis examined above is an attractive one, it presents a difficulty - and quite a considerable one. In fact, the measurement of the acceleration of the expansion *via* the Hubble diagram is the measurement of the quantity of dark energy responsible for that acceleration: it is the value of the cosmological parameter Ω_Λ. What is more, even in the absence of a unified theory of General Relativity and quantum mechanics, physicists can estimate the energy of the 'quantum vacuum' by way of simple considerations. In effect, such a theory brings in simultaneously the fundamental constants of relativity (G, the gravitational constant, and c, the speed of light), and the constant of quantum physics h (Planck's constant). On the basis of these constants, we can define a time t_{Planck} as well as a density ρ_{Planck} which we identify as that of the quantum vacuum. Sadly, the theoretical value and the measured value differ by some 120 orders of magnitude![8]

This difference between prediction and observation is, of course, enormous, and if we proceed along the same lines of reasoning in the case of the time other than the Planck time, there is a similar lack of agreement. Faced with these difficulties, we can, and indeed must, as the scientific method demands, seek out other possible hypotheses to ascertain whether they fit more successfully.

From 'quintessence' to a dark matter-energy

An elegant way of resolving the gap between the value of the density of the vacuum energy of the primordial universe and the value of the 'dark energy' currently present is of course to imagine that it could decrease with time, as with the density of ordinary energy. In a way, this is a hypothesis involving a variable 'cosmological constant'. What might seem here to be sleight-of-hand is in fact quite legitimate in physics. It is the basis of the so-called *quintessence* models, which rely on totally new concepts by introducing extra physical fields over and above traditional fields such as electromagnetic, gravitational, etc.

---

8. 1 followed by 120 zeros...

By definition, and without going into detail, a cosmological fluid of this type has a time-dependent equation of state and we appreciate the importance to physicists of the determination of the parameter *w* which characterizes such an equation. The birth of a new kind of physics may well depend upon this.

Another idea is to try to resolve at one and the same time the two big and troublesome questions of today's cosmology: dark matter and dark energy. It will be seen in fact that not only do these two components dominate the universe (by a long way), but that they are also in more or less the same proportions (approximately 1/3, 2/3). Also, we talk about the 'problem' of coincidence. Why do these two quantities not have very different values, if they are in no way related? It therefore seems tempting to imagine that we are dealing here with a single 'fluid' which might change its nature during the history of the cosmos. Several hypotheses have been put forward by theoreticians, for example, the idea of a new family of neutrinos whose mass would depend on their density in the universe. These *MaVaNs* (Mass Varying Neutrinos) would conveniently behave like a fluid at negative pressure, i.e. be capable of engendering cosmic acceleration while at the same time adding their contribution to dark matter.

Evolving constants?

Among other approaches, let us finally mention the possibility of some variation in the fundamental constants of physics: G, h and c.[9] If the speed of light had been greater in the past, this could be interpreted as an effect of greater distance. As the speed of light c intervenes in another aspect of physics *via* the *fine-structure constant,*[10] we need to carry out observations of different cosmic times (i.e. at different redshifts) within which the fine-structure constant plays a key role. We can therefore directly test the 'variation' of such a constant.

Doubtless, this will involve very delicate measurements, and even if our observations possibly indicate some variation in the fine-structure constant α (a subject of keen debate), the associated variation of c is too small to cause cosmic acceleration without also modifying the theory of gravitation.

As we see, the scientific problem described in this chapter still represents a considerable challenge, at both the observational and the theoretical levels. Measurements must be further refined, and our theories must delve deeper. A vast undertaking, but the stakes are considerable and the enthusiasm of cosmologists of all persuasions is running high. We are dealing here with no more and no less than the knowledge of the ultimate fate of our universe.

---

9. See Jean-Philippe Uzan & Bénédicte Leclercq: *The Natural Laws of The Universe: Understanding Fundamental Constants*, published by Springer-Praxis, 2008.
10. In atomic physics, the fine-structure constant α equals $e^2/\hbar c$.

Black is black

As we come to the end of this story, we ask one last question which perhaps some readers have already asked themselves: 'What is the purpose of all these efforts?'.

There is no one answer to this; it depends largely upon our opinions. For the most part, and apart from the obvious aim of furthering human knowledge in general and our understanding of fundamental physics in particular, the object of these efforts is also to discover what we can about the origin of our universe and thereby to be able to speculate about its future, even though this involves unimaginable timescales, over and above the duration of the human story itself.

The key parameter with which to end our story is naturally the total content of energy-matter, Ω_{tot}. If this parameter were definitely greater than 1, the universe would end by contracting towards a Big Crunch, a kind of anti-Big Bang. However, the verdict of the measurements is, as we have just seen, incontrovertible: Ω_{tot} seems to be well and truly equal to 1, and the expansion of our universe should continue into eternity, and even accelerate.

How will it all end?

Astrophysics tells us that the lifetime of our Sun will be no longer than 5 billion years. Now, a little speculation: the first significant phenomenon which should occur is the end of the stellar phase. As we have seen, a star is a self-gravitating celestial body of sufficient mass to trigger nuclear fusion reactions. During its life, part of the matter is re-injected into the surrounding medium, especially during the late stages of its evolution. Low-mass stars will generate planetary nebulae, and high-mass stars may become supernovae. Other matter will take the form of compact remnants such as white dwarfs, neutron stars or black holes.

As time passes, the composition of the gas in galaxies will therefore evolve, and the hydrogen will disappear to be replaced by more and more complex elements produced by stars during their lifetimes. When the hydrogen has practically disappeared, helium stars will replace them. The fusion of helium is, however, a less efficient process than the fusion of hydrogen, and helium stars will have short lives. And so it will go on, with the appearance of other stars made of more and more complex elements, stars which will be less and less efficient.

Finally, new stellar generations will no longer be guaranteed, and the stars will gradually die out one by one without being replaced. A few hundred billion years from now, even the least massive stars will have gone out, and the stellar phase of the universe will be finally at an end. The galaxies will no longer contain any luminous objects, but merely stellar remnants such as white dwarfs, neutron stars or black holes, with associated planets. The universe would then be devoid of all sources of visible light and would appear dark to the eyes of our very distant descendants.

The next stage will see the evaporation of the galaxies, a process which will take perhaps a billion billion years. In time, as a result of complex mechanisms of interaction, every galaxy will gradually have lost 99 per cent of its star remnants.

Simultaneously, the central density of the galaxies will have increased towards a critical value beyond which the galactic centre will have been transformed into a black hole.

The same process of evaporation and transformation into a black hole will then occur for whole clusters of galaxies, over an even longer period of time. Once this process has been concluded, galaxies and clusters of galaxies will be just a very distant memory. Our universe will have been transformed into a multitude of black holes of different masses, together with neutron stars, white dwarfs which have grown dark, and planets dispersed among these objects.

But there might be even worse to come, in an even more radical scenario...

Some theoreticians think that the '(anti-) gravitational' action of dark energy could become preponderant on a local scale, even down to the subatomic and nuclear scales. All objects, from the cosmos itself down to the nuclei of atoms, would literally fly apart, and no binding force would be capable of resisting their massive destruction. There are some who have even drawn up a (still very uncertain) calendar showing the inexorable 'countdown'. This apocalyptic schedule suggests that the 'end of ends' will occur in only 20 billion years from now. In about 19 billion years, the expansion of the universe will have caused all the other galaxies to be no longer visible from our own. The sky will indeed be a desolate place in those days. The Milky Way will begin to break up 60 million years before the 'end of ends', and only 3 short months will remain before our solar system begins to disintegrate in its turn.

Whatever the real future of the universe might be, it certainly seems that it will be a cold and dark one.

From false to true...

As in all good scenarios, there is perhaps a way out. If the dark energy responsible for this somber fate is indeed a 'vestige' of the energy of the quantum vacuum of the primordial universe, there is nothing to prove that this 'vacuum' is a 'true vacuum'. In simpler terms, the universe was perhaps not in its most fundamental state at that time.

We know that a pencil can balance on its end, but we also know that this is an unstable configuration. Just like the pencil, the universe is perhaps in a 'false-vacuum' state, and could at any moment 'topple' into another state. Then, everything would have to be rewritten, and physics (or at least fundamental constants) would be different.

But that is another story...

Appendices

These appendices are aimed at the more informed reader who wishes to investigate certain aspects of this book in greater detail.

1. Hydrostatic equilibrium

A self-gravitating system may be in equilibrium if a compensating force is acting. So, in all cases, we can write what is known as the equation of *hydrostatic equilibrium*, which expresses the equality of the two forces.

Spiral galaxies

For a star of mass m_*, rotating at circular velocity v_* in a spiral galaxy, it is centrifugal force which must be considered.

Equilibrium gives:

$$\frac{GMm_*}{r^2} = m^* \frac{v_*^2}{r}$$

If we suppose that the total mass M is concentrated in the central hub of the galaxy, it follows that the velocity decreases with the distance according to Kepler's law:

Keplerian decrease $v_* = \sqrt{\dfrac{GM_*}{r}}$, contrary to the flat rotation curves observed.

If we then suppose the existence, in addition to the central stellar hub, of a spherical *dark halo*, whose density ρ varies as r^{-2}, we can show that the contribution of this halo of mass $M_{halo} = \frac{4\pi r}{3}$ to the total velocity will be a constant, beyond a sufficiently large distance from the centre:

$$v_{Halo} = \sqrt{\frac{GM_{Halo}}{r}} = \sqrt{\frac{4\pi G}{3}}$$

Figure A1 illustrates the application of this principle to a real galaxy, and the actual contributions of the various components to the ordinary matter of the galaxy (the disc and hub stars and the interstellar gas), to which is added the contribution of the halo of dark matter. We see that it is perfectly possible to give an account of the measurements.

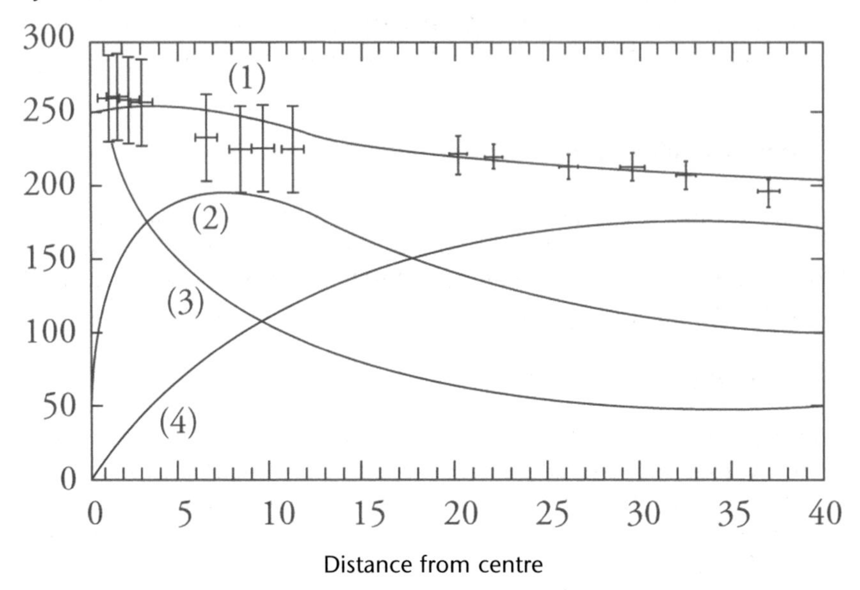

Figure A1. Curve (1) is the curve adjusting the observations given by the measured velocities and their error bars. Curve (2) corresponds to the contribution of the galactic disc, and curve (3) to the 'Keplerian' contribution of the central hub of stars. Curve (4) is the contribution of the spherical dark halo.

Clusters of galaxies

The equation of hydrostatic equilibrium is an equation involving local quantities: the pressure gradient, and gravitational force. It is often very useful to have at one's disposal an expression for this condition of equilibrium, but using global quantities instead. We therefore bring in two essential physical quantities: the *total kinetic energy* of the system E_c and the *gravitational potential energy* U_g, in the so-called Virial relationship

$$2E_c + U_g = 0$$

The Virial theorem may be applied to gravitational systems such as clusters of galaxies, and involves the relationships of the various observable quantities present: the positions and velocities of the objects in question. Taking into account the effects of projection on the sky, we can demonstrate that:

$M_{Viriel} = \frac{3\pi}{2G}\,\overline{V^2}\,\overline{R}$ where $\left[\overline{V^2}\right]^{1/2}$ is the *velocity dispersion of the galaxies* (i.e. the measurement of the agitation velocities of the individual galaxies and therefore of their kinetic energies) in the cluster, and $\overline{R}$ is a characteristic 'dimension' of the system (obtained for example by averaging out the positions of the galaxies as determined with a digital camera or photographic plate).

Taking as typical values 1000 km/s for the velocity dispersion and 1 Mpc for the dimension of the cluster, we obtain a dynamic mass of the order of 10^{15} solar masses, about 5 to 10 times greater than the luminous mass.

2. Matter in all its states

The four possible states of matter are: solid, liquid, gaseous and plasma. In the case of plasma, the matter is ionized, i.e. the electrons and the nuclei of atoms are no longer associated, although the matter remains globally neutral. We shall be discussing principally the gaseous and the plasma states here.

Perfection in gases

A solid exerts no pressure because the molecules are held together by the so-called (cohesive) Van der Waals forces. Gas exerts pressure, and we may remember from our schooldays the famous 'ideal (or perfect) gas equation' $PV = NkT$ relating the pressure P, volume V and temperature T of a gas. k is Boltzmann's constant, and N the total number of particles in the volume V.

Let us return for a moment to the way in which this *equation of state* is established. The pressure of a gas is the manifestation of the motion of the particles of which it is composed, which transfer their momentum (equal to the product of their mass and their velocity) to the walls of the container. This pressure is by definition a force per unit area of the surface. In dimensional terms, this amounts to an *energy density* ε, and it is this definition with which we shall proceed.

The kinetic energy (the product of the mass and the square of the velocity, divided by 2) to be considered for the total amount of gas is therefore N times that of one particle, from which we deduce, taking into account that all directions are equivalent, that:

$$P = \frac{2\varepsilon}{3}$$

where $\varepsilon = \frac{N}{V}\frac{mv^2}{2} = nm\frac{v^2}{2} = \rho\frac{v^2}{2}$ is the kinetic energy density of the gas.

This energy can also be measured, according to thermodynamics, by the temperature T, which leads to the (*thermal*) pressure of an ideal gas:

$$P_T = nkT \text{ and } P_T = 2kT\rho/m_p$$

for a hydrogen plasma where the number of protons of mass m_p is equal to the number of electrons.

A photon 'gas'

Whoever has observed an open fire or the glow of a cigarette may notice that, the higher the temperature T, the 'whiter' the radiation becomes (hence the expression 'white hot'). It may seem an easy undertaking, using classical physics, to explain how the intensity of the radiation in such a system depends on temperature. However, a challenge to this assumption led to one of the most important scientific revolutions of modern times, with the introduction of quantum mechanics.

According to this new discipline, electromagnetic radiation has a double aspect: it is at once wavelike and corpuscular. 'Light', for example, consists of *quanta of energy* hν (photons), ν being the frequency of the radiation, and h Planck's constant.

It can be shown that the energy density of such a 'photon gas' or 'black body' depends on the fourth power of the temperature! So, we can write

$$\varepsilon = aT^4, \text{ with } a = \frac{8\pi^5}{15}\frac{(kT)^4}{(hc)^3} = 7.55\ 10^{-16}\ J{\cdot}m^{-3}{\cdot}K^{-4}$$

The photons therefore exert a pressure (the so-called *radiation pressure*), which exhibits the same behavior:

$$P_r = \frac{aT^4}{3}$$

Degenerate matter

Not only does quantum mechanics dictate the discrete nature of energy and exchanges of energy, but it also affirms, *via* Heisenberg's Uncertainty Principle, that the motion of a particle restricted to a given space or during a given interval of time is subject to an irreducible uncertainty. Thus, if the available space is limited, in a certain direction, to Δx, the motion quantity $p = mv$ will be defined with uncertainty Δp such that:

$$\Delta p\ \Delta x \geqslant \hbar, \text{ where } \hbar = h/2\pi.$$

The consequences of such a relationship are fundamental. It means that a quantum particle in a 'box' will never have zero energy since there will always be a so-called *Fermi contribution* (ΔE_F), which is expressed thus:

$$\Delta E_F \sim (\Delta p)^2/2m, \text{ with } \Delta p \sim \hbar/\Delta x.$$

There is another essential consequence of this for a plasma, i.e. a collection of nuclei and electrons which, from a quantum point of view, are fermions. These fermions must obey another quantum rule which stipulates that two identical fermions cannot be in the same quantum state: this is known as the Pauli Exclusion Principle. In this notion of the 'quantum state', spatial position is one element.

Consequently, with all the other physical parameters being identical, two fermions cannot occupy the same spatial volume.

As a first approximation we deduce therefore that for N fermions of mass m_F contained in volume V, each one will have 'access' to only one individual volume $\Delta V \sim V/N$, whence, *via* the *Fermi energy density* of these N fermions:

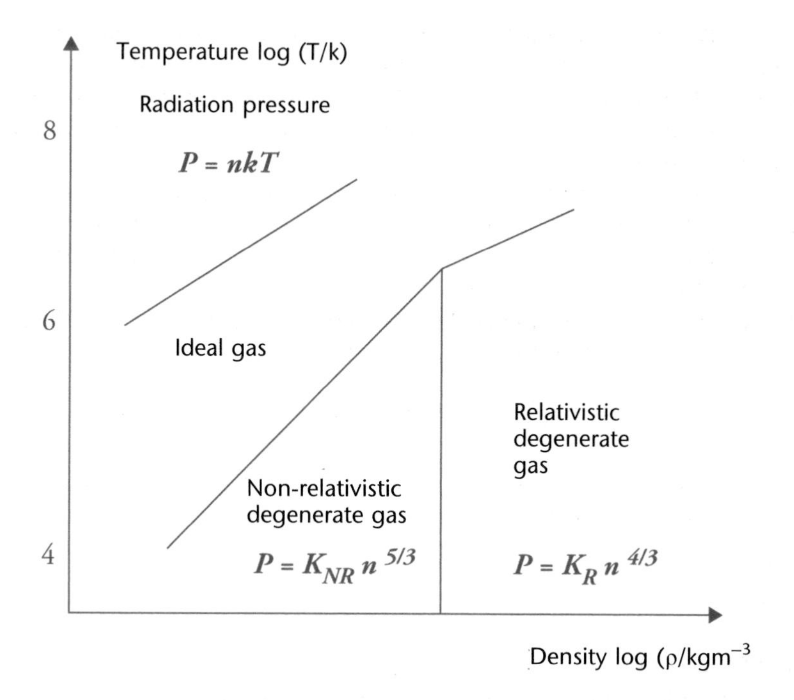

Figure A2. Temperature – density diagram. Different types of stars are located on this diagram according to conditions of temperature and density which imply different equations of state. Solar-type stars are in the zone marked 'ideal gas'; 'massive' stars are in the zone marked 'radiation pressure'; and finally white dwarfs are in the zone marked 'non-relativistic degenerate gas'.

$$\left(\varepsilon \sim \frac{1}{V}\, N\Delta E \sim \left(\frac{N}{V}\right)^{5/3}\left(\frac{\hbar^2}{2m_F}\right) \sim (n)^{5/3}\left(\frac{\hbar^2}{2m_F}\right)\right)$$

and, by a more complete calculation, the expression of the *Fermi pressure* or *quantum pressure* for the electrons:

$$P_F = K_{NR}\,(n_e)^{5/3}$$

with $K_{NR} = \left(\frac{3}{8\pi}\right)^{2/3} \frac{h^2}{5m_e} = \sim 10^{-37}$, which does not depend on temperature and is inversely proportional to the mass of the fermion in question. We can therefore summarize the different states of matter and the different types of 'possible' stars in a 'temperature-density' diagram, as in Figure A2.

3. Star profile

Identification checklist

A star is characterized by six quantities:

- mass M_*;
- luminosity L_*, corresponding to the energy emitted per second;
- radius R_*;
- effective temperature T_e, defined as the temperature of a black body emitting the same quantity of energy per surface area, or $L_* = 4\pi R_*^2 \sigma T_e^4$;
- age t_*;
- chemical composition (X, Y, Z), where X, Y and Z are the proportions by mass of hydrogen, helium and metals.

In the case of the Sun:

$M_\odot = 2 \cdot 10^{30}$ kg;
$L_\odot \sim 4 \cdot 10^{26}$ W;
$R_\odot = 7 \cdot 10^{8}$ m;
$T_e \sim 5\,800$ K;
$t_\odot \sim 10^{10}$ yrs;
$X_\odot \sim 0.73$;
$Y_\odot \sim 0.25$;
$Z_\odot \sim 0.02$.

These different parameters are not all independent. There are two principal relationships: luminosity and effective temperature ($L \sim T_e^{6.7}$); and luminosity and mass ($L \sim M^4$) for masses $\sim$ the mass of the Sun.

The first relationship corresponds to the *H-R Diagram* (*Hertzsprung-Russell* or *color-magnitude diagram*), on which the majority of stars are arranged in four groups.

The main group contains 90 per cent of the stars and corresponds to the Main Sequence, upon which the stars spend most of their lives, burning hydrogen. The other groups are the giants and supergiants which lie to the upper right of the H-R Diagram and the white dwarfs which lie to the lower left.

Questions of gradients

We have already examined the equilibrium of systems of 'particles' in the wider sense of the term: galaxies consisting of stars; and clusters of galaxies.

In the case of a plasma as found in the interiors of stars, we can describe the equilibrium between gravity and the forces of pressure in two ways.

First, in a global context, involving the Virial theorem relating internal energy E to gravitational energy U_g. Then, in a local context, pressure and gravitational force must equal each other at all points along any radius, assuming a spherical structure.

We saw earlier that pressure can take different forms (thermal, radiation, Fermi), each characterized by a specific equation of state. The total pressure is written thus:

$$P = P_{th} + P_{rad} + P_F$$

From the condition of hydrostatic equilibrium, we can easily deduce an expression for the mean value of the pressure of a star:

$$\bar{P} \sim 0.3G\rho^{4/3} M^{2/3}$$

In the case where the thermal pressure is the dominant contribution, we obtain estimates for the mean density $\bar{\rho} = \frac{3M}{4\pi R^3}$ and the mean temperature $\bar{T} \approx 0.1 \frac{Gm_p}{k} \frac{M}{R}$.

Estimates for the Sun are:

$$\bar{\rho}_{\odot} \sim 1.4 \times 10^3 \text{ kg m}^{-3} \text{ and } T_{\odot} \sim 2.4 \times 10^6 \text{ K}$$

Also, a detailed model of the Sun shows that the central density ρ_c is about 100 times the mean density, and the central temperature T_c is about 4 times the mean temperature. We can compare the mean thermal pressure with the radiation pressure:

$$\frac{P_r}{P_{th}} \approx \frac{aT^4 / 3}{2kT\rho / m_p}$$

We obtain, for one solar mass, a value for this relationship of $\sim 10^{-4}$, showing that it is negligible for such a mass. However, since this varies as the square of the mass:

$$\left(\frac{P_r}{P_{th}} \propto \frac{T^3}{\rho} \propto \frac{M^3 / R^3}{M / R^3} \propto M^2 \right)$$

We see that radiation pressure will be important in high-mass stars and will dominate for $M \sim 100$ solar masses or greater.

As for the Fermi pressure for electrons, compared with the central thermal pressure:

$$\frac{P_F}{P_{th}} = \frac{K_{NR} \left(\frac{\rho}{2m_p} \right)^{5/3}}{2kT\rho / m_p} \cong 10^{-1}$$

we obtain a value of the order of 10^{-1} for the ratio of the two quantities, confirming that the degeneracy pressure is negligible in the normal conditions of the Sun. Writing the ratio of thermal pressure and relativistic Fermi pressure:

$$\frac{P_{FR}}{P_{th}} = \frac{K_R \left(\frac{\rho}{2m_p} \right)^{4/3}}{GM^{2/3} \rho^{4/3}} \cong \frac{K_R}{(2m_p)^{4/3} GM^{2/3}}$$

we see that this degeneracy pressure will prevail when the mass of the core of the star is of the order of:

$$M \cong \left(\frac{K_R}{G} \right)^{3/2} \frac{1}{(2m_p)^2} \sim 10^{30} \text{ kg} \sim 1 M_{\odot}$$

approaching the 'canonical' value of Chandrasekhar's mass $M_{Ch} \sim 1.44 M_{\odot}$.

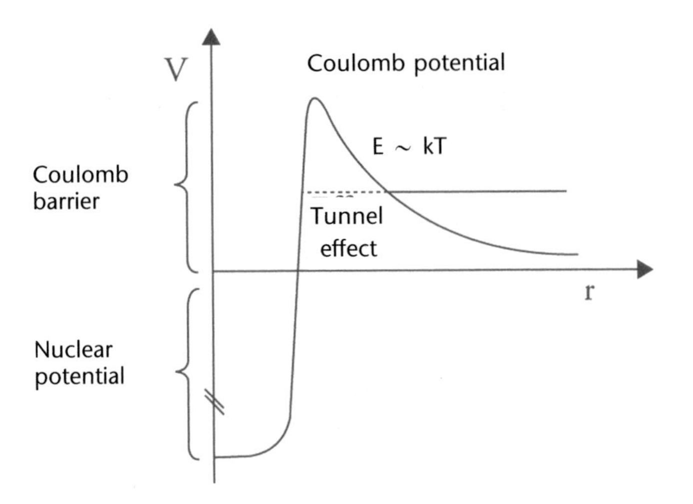

Figure A3. Two 'classic' particles with the same electric charge cannot overcome the Coulomb potential barrier in order to 'reach' nuclear potential in the energy conditions existing in stellar cores. The 'quantum tunneling effect', predicted by quantum mechanics, allows penetration at lesser energies, thereby making fusion reactions possible.

4. The quantum tunneling effect

A barrier to be crossed

Two positively charged nuclei have to overcome the Coulomb barrier, which varies as $1/r^2$ (Figure A3), to achieve separations at which the strong interactions will take over ($\sim 10^{-15}$ m = 1 fermi). To overcome this repulsive barrier, classical mechanics teaches that more and more kinetic energy is required as the distance between the particles closes, making interaction very difficult, to the point of improbability. In fact, the height E_c of this barrier gives, for nuclei of charges Z_1 and Z_2, separated by distance r:

$$\frac{Z_1 Z_2 e^2}{4\pi\varepsilon_0 r} = \frac{\alpha Z_1 Z_2 \hbar c}{r}$$

(where $\alpha \cong 1/137$ is the so-called fine-structure constant).

So, we have Ec (MeV) $\sim Z_1\, Z_2/r$, with r in fermi.

To attain 1 fermi, for example, the energy required is about 1 MeV, while the energy available in a star at 10 million K is a thousand times less, equal to 1 keV (in effect kT (keV) $\sim 10^{-7} T$ (K)). Although, in a plasma, the energies of the particles, subject to the 'Maxwell-Boltzmann' law, can attain values exceeding this mean value, only a fraction ($\sim 10^{-430}$) would have sufficient energy, while the number of atoms of hydrogen in the Sun is only 10^{57}!

However, in quantum mechanics, the Heisenberg uncertainty relation, here expressed for energy and time: $\Delta E \Delta t \geqslant h$, allows the possibility of the appearance

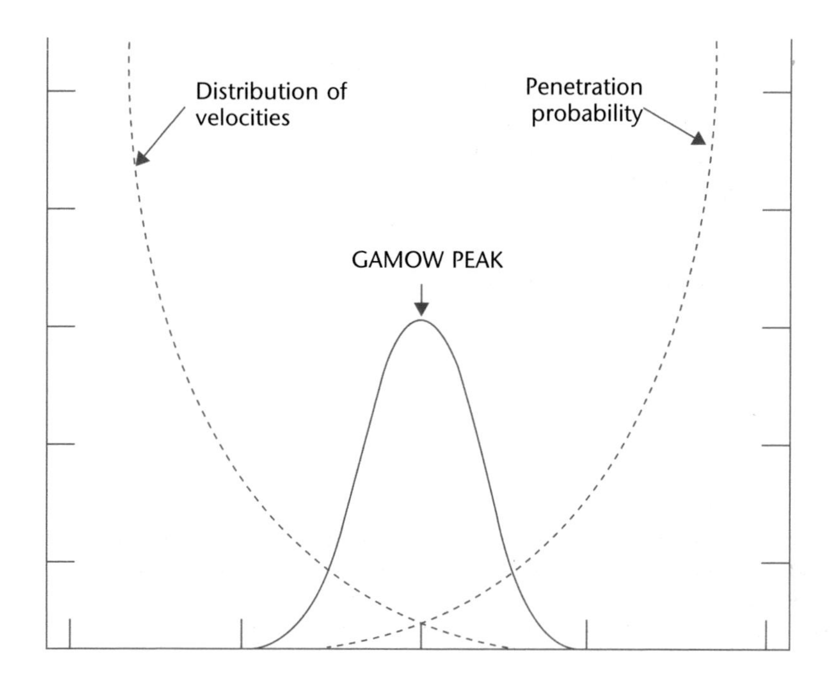

Figure A4. In order to interact, nuclei at temperature *T*, following the law of the Maxwellian distribution of velocities, must cross the Coulomb barrier separating them. This is made possible, with a certain probability of penetration, by the 'quantum tunneling effect'. The Gamow peak results from the product of this Maxwellian distribution of velocities and the probability of the quantum tunneling effect. It corresponds to the energy at which the reaction will take place.

of a supplementary energy fluctuation ΔE of duration Δt. This notion is translated into reality through the existence of the *tunnel effect*, discovered by physicist George Gamow during his work on radioactivity. In the context of this quantum effect, a particle is characterised by a *penetration probability* $P \sim e^{-(E_G/E)^{1/2}}$ where E_G (the *Gamow energy*) intervenes (in the case of two protons, $E_G = 2mc^2 (\alpha\pi Z_1 Z_2)^2 \sim 500$ keV).

It follows from this that, contrary to the classical case, a proton possesses a non-zero probability of crossing the Coulomb barrier and interacting with another proton *via* the strong interaction.

The probability that the fusion of these two nuclei will occur is consequently given by the product of the penetration probability and the Maxwell distribution law:

$$P \sim e^{-(E_G/E)^{1/2}} e^{-E/(kT)},$$

– a product which will produce a maximum (Figure A4) at the *Gamow peak* (of energy $E_{PG} \sim ([E_G(kT)^2]/4)^{1/3}$ and width ΔE_{PG}).

In the case of two protons and typical energy $kT \sim 1$ keV, we therefore have:

$E_G \sim 500$ keV, with the Gamow Peak at $E_{PG} \sim 5$ keV, with a width of $\Delta E_{PG} \sim 5$ keV, which now makes the reaction probable. It becomes more and more improbable at lower temperatures, i.e. at lower masses, which fixes the lower limit for a star at ~ 0.1–0.2 solar masses.

5. Fusion, fission and stellar lifetimes

$E = mc^2$!

Atomic nuclei consist of protons (positively-charged particles) and neutrons (neutral) bound together by the strong force. The mass of a nucleus does not equal the mass of its nucleons. It is therefore theoretically possible to 'recover' the corresponding energy (*via* the relation $E = mc^2$) by forming a new element from the available nucleons.

The difference in masses represents the *binding energy* of the nucleus in question. Consequently, a nucleus composed of Z protons of mass m_p and N neutrons of mass m_n will have a binding energy:

$$E_L\,(Z, N) = [Zm_p + Nm_n - m\,(Z, N)]\; e^2$$

In Figure A5, we see that the energy of the nucleon, defined as $E_L\,(Z, N)/A$, increases with the atomic mass A as far as iron ($A = 56$), and that this increase is followed by a slow decrease.

This increase is considerable from hydrogen to helium-4. Consequently, the fusion of hydrogen to form helium will liberate far more energy per unit of mass than, for example, the fusion of helium to form carbon.

Energy is therefore released if light elements fuse to form heavier elements, up to and including iron. Beyond this, fission processes, with heavier elements dividing to form lighter elements, will liberate energy.

A long-lived Sun

With this information, we can now estimate the lifetime of the Sun.

We know that four atoms of hydrogen can fuse to produce a helium atom. The mass of helium thus formed is 0.7 per cent lighter than the sum of the masses of the individual nucleons composing it (the mass of helium He-4 is 3.9726 times that of hydrogen (m_p)). This difference in mass represents the radiative energy E. If we consider that only 10 per cent (the core) of the total mass of the Sun is affected by this process, we obtain:

$$E \sim 0.1 \times 0.007 \times M_{\odot}\, c^2 \sim 10^{44}\ \mathrm{J}$$

Knowing that the Sun radiates, per unit of time $L_{\odot} = dE/dt = 4 \cdot 10^{26}$ W, the lifetime of the Sun is at least:

$$T_{nuc} \sim E/(dE/dt) \sim 10^{44}/4 \cdot 10^{26} \sim 2.5 \cdot 10^{17}\ \mathrm{s} \sim 10 \text{ billion years.}$$

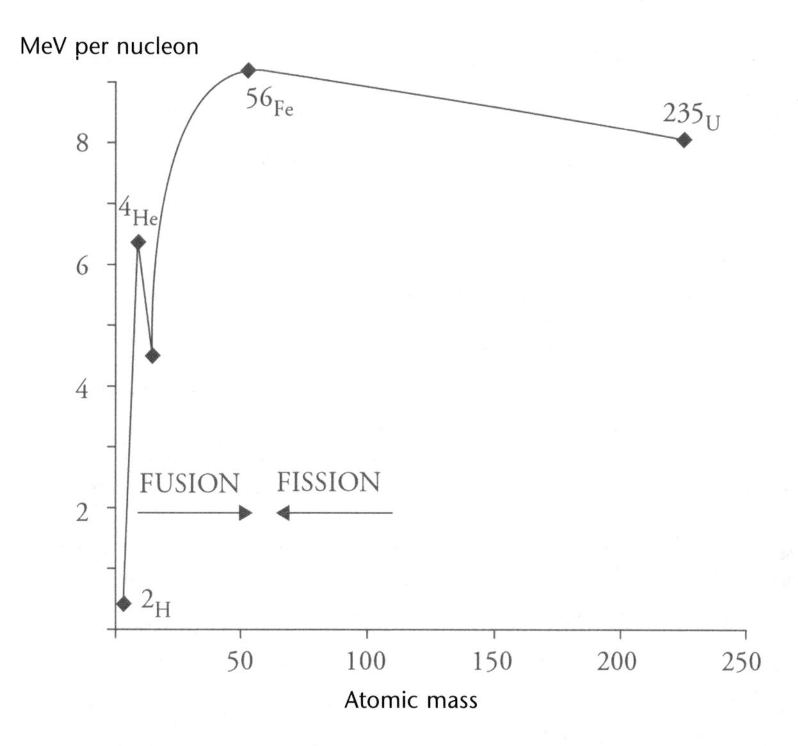

Figure A5. Nucleon binding energy as a function of atomic mass.

This is far longer than the *Kelvin-Helmholtz* time t_{KH}, based upon the transformation of gravitational potential energy $U_g \sim G(M_\odot{}^2/R_\odot)$ into radiation:

$t_{KH} \sim U_g/L_\odot \sim 10^{41}/4 \cdot 10^{26} \sim 10$ million years.

6. Gifts from the stars

What is the origin of the elements that surround us in our world of living and nonliving matter? They are the products of *nucleosynthesis*, which is defined as the ensemble of nuclear processes which create the various chemical elements as classified on the periodic table of Mendeleev.

Within the context of the study of the cosmos, we can identify four types of reactions and four phases corresponding to various astrophysically important locations and different periods of cosmic history.

I) The first phase is that of the *primordial nucleosynthesis* which occurred during the first three minutes of the universe, and during which *thermonuclear fusion* proceeded. In this reaction, the mass of the progenitor nuclei was greater than that of the nuclei produced and the excess mass provided the energy of the stars ($E = mc^2$). The elements formed were hydrogen, deuterium and the two stable isotopes of helium (He-3 and He-4), together with traces of lithium and beryllium.

II) The second phase took place at the hearts of stars *(stellar nucleosynthesis)*, where, in similar fashion, fusion reactions occurred while the stars remained in equilibrium for millions, even billions of years. The elements formed populate Mendeleev's periodic table up to and including iron.

III) As we have seen, stars of sufficient mass end their lives in an *explosive phase* (e.g. supernovae) during which new elements are formed.

Phenomena such as *neutron capture* and *photodisintegration* take place as the explosion occurs. The neutrons produced by fusion reactions have a lifetime of about 15 minutes during which they may be captured by nuclei to form other, heavier elements. Also, the temperature may become so high that photons will have enough energy to break the heavy nuclei, to produce lighter ones. The elements formed are those beyond the iron peak, and also some lighter elements.

IV) Several generations of stars live out their lives, gradually enriching the interstellar medium with heavy elements. The elements of this medium are continuously bombarded by these very energetic nuclei from explosions of other stars, forming what is known as *cosmic radiation*.

Spallation reactions occur, during which cosmic radiation may break heavy nuclei in the interstellar medium to create lithium, beryllium and boron.

So, billions of years on, the cycle is complete, and all the necessary elements for the chemistry of life have been created.

7. The Roche lobe

The notion of the *Roche lobe* stems from the work of Edouard Albert Roche, and was largely based on the findings of Joseph-Louis Lagrange. In the course of his research on celestial mechanics, Lagrange had drawn attention to particular locations, known as 'Lagrangian points'. At these points, the force of gravity of two bodies balances out, and a third body is able to remain there in a state of equilibrium. This situation is completely independent of the nature of the bodies, which may be stars, planets or even an artificial satellite. Only the masses and the distances involved are relevant.

Edouard Roche completed Lagrange's work by calculating the form of the equipotential surfaces around each of the celestial bodies, i.e. the places where the force of gravity has the same value.

Around a single massive body, these surfaces are spherical. In a binary system, they form two families centred on each of the bodies and assume, with distance, a characteristic elongated 'teardrop' shape.

The Roche lobe is that region of space where particles are gravitationally in

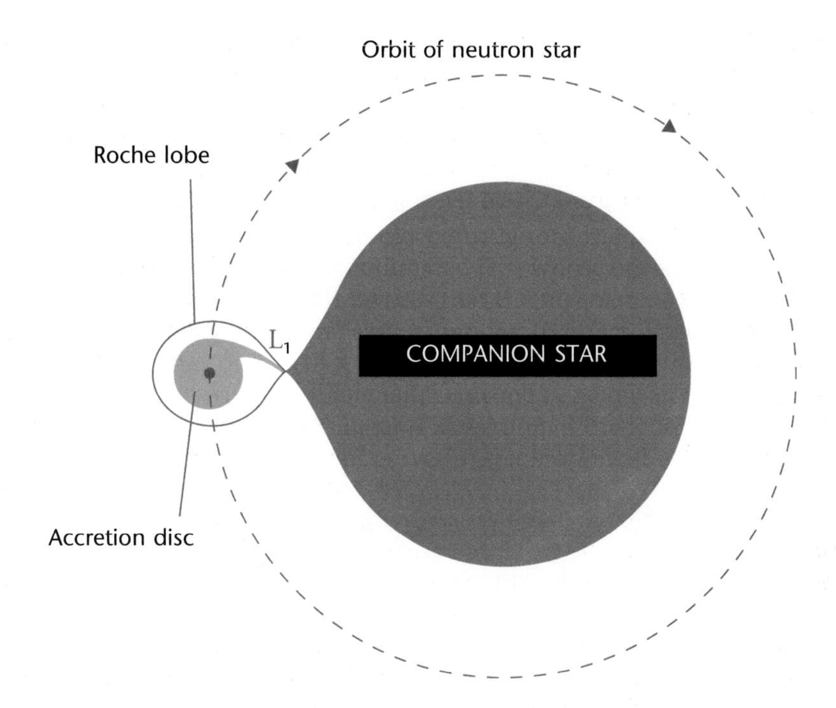

Figure A6. The Roche Lobe is that region of space where particles are gravitationally in equilibrium between the two stars of a binary system. Mass transfer from the companion star (for example, a red giant) towards the orbiting neutron star is possible *via* the Lagrangian point L_1. This matter is accreted by the compact object, whose mass increases (which may cause it to explode).

equilibrium between the two bodies (equipotential zero). It passes through the well known first Lagrangian point L_1 (Figure A6).

So, in the case of binary stars, where one of the bodies is a red giant star and the other is a white dwarf, the atmosphere of the red giant may extend beyond the Roche lobe and be gravitationally captured by the companion star. This may lead to the complete disintegration of the first star, with the companion acting as a kind of 'vacuum cleaner'. The matter which falls onto the companion star 'feeds' it and leads to 'cataclysmic' phenomena.

8. A revealing radiation

Whatever the exact nature of the mechanism responsible for the 'central engine' of gamma-ray bursts, the physical context is indeed extreme. We have to imagine it as a very hot, very dense and chaotic plasma, within which a strong magnetic field is generated.

Charged particles, and especially electrons, will acquire a circular velocity because of this field, and spiral in the direction of the magnetic field. The circular frequency of an electron, in this relativistic case and for a magnetic field of intensity B, is:

$$\nu_c = \omega_c/2\pi = eB/(2\pi\gamma m_e).$$

However, being accelerated (centripetal acceleration due to rotation), these particles will emit electromagnetic radiation, according to Maxwell's laws of electromagnetism, and, depending on whether or not their velocities v are relativistic, they will emit either *cyclotron* or *synchrotron* radiation.

The criterion separating these two regimes is given by the value of the Lorentz factor Γ:

$$\Gamma = \frac{1}{\sqrt{1 - v^2 / c^2}}$$

If $\Gamma \sim 1$ ($v \ll c$), the radiation will be of the cyclotron type. If $\Gamma \gg 1$, it will be of the synchrotron type.

The synchrotron radiation involving the relativistic particles possesses two characteristics.

The first corresponds to the fact that this radiation is emitted only in a limited region of space (*beam collimation*), i.e. in a cone of opening angle θ such that:

$$\theta \sim 1/\Gamma$$

i.e. the cone becomes narrower as the particles become more relativistic (Figure A7). To detect them, an observer must be in the line of sight.

The other characteristic of synchrotron radiation is that its spectrum, i.e. the intensity as a function of frequency $F(\nu)$, is in the form of a power law.

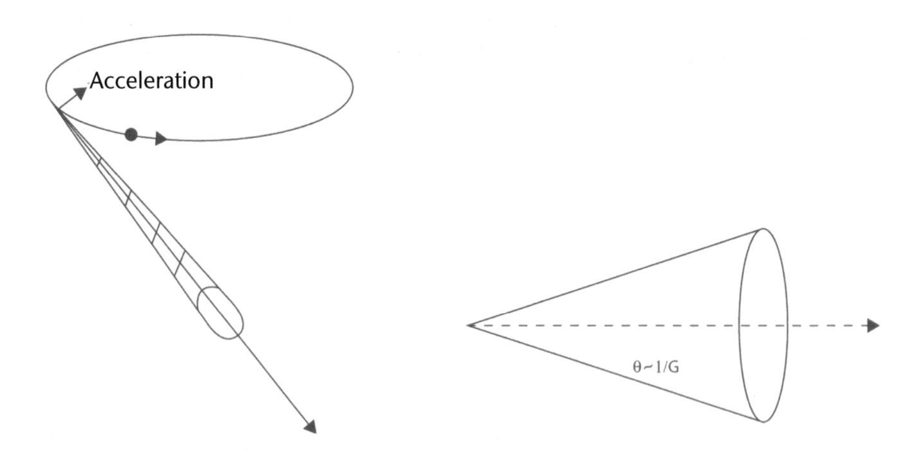

Figure A7. Synchrotron radiation is emitted in a cone whose opening angle depends on the Lorentz factor.

Indeed, an individual electron moving helically in the magnetic field will emit radiation of typical frequency ν_c. However, we need to take into account the entirety of the particles in the population and its energy distribution. In the case of particles accelerated by the passage of successive shock waves (see below), the distribution is no longer Maxwellian but becomes a power law:

$$N(E)\mathrm{d}E \sim E^{-p}\mathrm{d}E$$

The result for the spectrum of the radiation is therefore similarly a power law of slope $s = (p - 1)/2$ where p is the index of energy distribution of the electrons.

Finally, various processes such as the reabsorption of the synchrotron radiation, or the fact that the electrons lose energy rapidly *via* their own emissions, mean that the synchrotron spectrum will exhibit a 'break' above a certain energy.

The spectra of GRBs also shows this break, which becomes a maximum known as E_{peak} when we consider the quantity $E^2N(E)$.

9. Waves and shocks

Sound waves

As an aircraft moves through the air, it momentarily displaces air molecules, continuously creating sound waves. These sound waves correspond to successions of compression and rarefaction and propagate in all material media. They move out from the aircraft symmetrically, like ripples on a lake, formed by an object falling into the water (Figure A8).

The velocity c_s of the waves is determined by the medium (and the physical conditions) through which they pass. The speed of sound in a liquid or a solid is, for example, greater than it is through air.

In the case of the aircraft, it forces a passage all along its path and, after a time interval t, all the molecules within the sphere of radius $c_s\ t$, centred on the

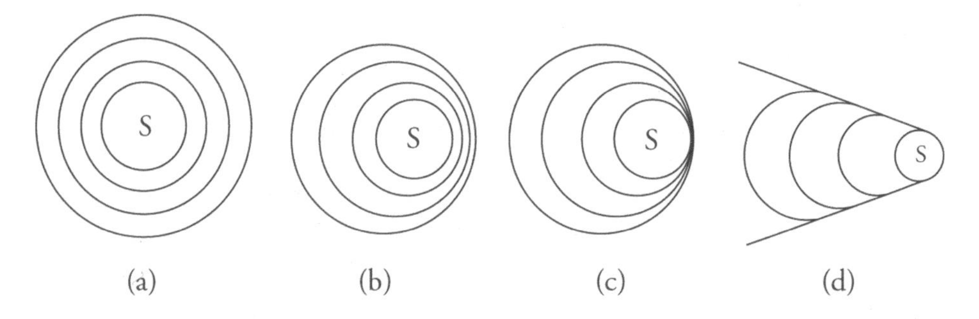

Figure A8. (a) Wave emitted by a motionless source S. **(b)** Wave emitted by a source S moving at a velocity less than that of the waves. **(c)** Wave emitted by a source S moving at the same velocity as that of the waves. **(d)** Wave emitted by a source S at a velocity greater than that of the waves.

aircraft, have been 'informed' of its passage. Meanwhile, the aircraft moves on, and the successively created spheres remain contained one within another, all the while the aircraft is moving slower than the waves created.

When the aircraft itself reaches the speed of sound, its speed is by definition Mach 1.

The shock wave

The aircraft may now move even faster than the waves, continuing to create them in front of it, and these waves will still propagate at a speed dictated by the physical conditions of the medium involved. It can therefore overtake, at a given moment, the wave which it has just created. The spheres then intersect, with superposition of physical effects (including pressure) at the points of intersection. As this phenomenon of wave emission and sphere creation is continuous, the points of superposition form a continuous surface (a 'caustic' or 'envelope'), in the shape of a so-called 'Mach cone' with the aircraft at its apex.

In front of the caustic, the medium is not yet 'informed' of the arrival of the wave, while at its surface, the pressure variation has been amplified. A discontinuity has been created, which can be represented schematically as a step on a staircase: the *shock wave*.

In a fluid, the shock wave is therefore the site of abrupt modifications in physical properties such as speed, pressure and temperature. This may lead to phenomena which are particularly violent at the moment of occurrence, for example the sonic 'boom' made by an aircraft flying at supersonic speed (Figure A9). There may also be an important gain in energy for particles involved in a shock wave: a kind of continuous 'ping-pong' effect as the particles collide with the shock front and their energy is augmented by the *Fermi process*. The particles may then re-emit the energy in the form of very energetic photons.

It is this kind of phenomenon which occurs in the regions of the jets expelled by very active celestial bodies.

A light-shock!

Nothing travels faster than light... in a vacuum! However, in a material medium, its speed decreases by a factor known as the 'refractive index' of that medium. The speed of light in a given medium, if n is the refractive index of that medium, is $c' = c/n$. In water, for example, the speed of light is $0.75 \cdot c$. If a charged particle moves in a given medium, there will therefore be, *via* electromagnetic interaction, the emission of photons propagating at a speed of c'. If the particle responsible for this emission moves at velocity v greater than c', an effect analogous to the one we have just discussed will occur. The radiation emitted (known as 'Cherenkov radiation') is limited to a cone similar to the Mach cone.

Figure A9. An American fighter plane breaks the sound barrier over the Pacific (US Navy).

10. Measurements and distances

Events and measurement

In classical physics, space is limitless, absolute and rigid, and exists *en soi* (that which exists in itself), independently of the physical phenomena which occur in it. We can envisage it as a stage upon which every phenomenon is located by its coordinates, for example x, y, and z, in the three-dimensional space. The notion of coordinates means that we can define a distance (or spatial *metric*) between two points. So the distance $\mathrm{d}r$ between two points is: $\mathrm{d}r^2 = \mathrm{d}x^2 + \mathrm{d}y^2 + \mathrm{d}z^2$, where $\mathrm{d}x$, $\mathrm{d}y$ and $\mathrm{d}z$ are the differences in the coordinates between the two points.

In this kind of physics, time is also an absolute. It is moreover a separate notion from that of space. It 'flows' uniformly and seems to be a parameter for the classification of phenomena. In relativity, time and space are inseparable, and it is no longer sufficient to talk of positions in space. We therefore introduce the idea of *events*, while the notion of distance is generalized. Special relativity, which can be seen as a particular case of general relativity in which gravity is absent and space remains flat, stipulates that speed of light is a constant c, whatever the respective motions of the source and the observer. Here, the distance $\mathrm{d}s$ between two events is given by:

$$ds^2 = c^2\, dt^2 - dr^2$$

where d*r* is the spatial distance and d*t* the difference in time between the two events. We see that (sign excepted) time (multiplied by c in order to obtain a length) is involved as spatial coordinates are, and becomes a fourth dimension. So we have our definition of a space-time metric.

The relative notion of distances

In general relativity, space is 'curved' by its energy-matter content and the most general metric assumes complex forms. Fortunately, in the context of cosmology, the Cosmological Principle brings considerable simplifications, especially in the question of the existence of *cosmic time.*

The metric of a homogenous and isotropic universe now takes the form:

$$ds^2 = c^2 dt^2 - R(t)^2 dr^2$$

where R(t) is a function of time. At a given moment ($dt = 0$) the metric reduces to $R(t)^2 dr^2$ and provides a measurement of distances.

We see that, in time, these distances are modified by the function $R(t)$, which therefore has the role of a scale factor. The distances are multiplied by $R(t)$, and the volumes by $R(t)^3$. It is the scale factor $R(t)$ which takes account of the expansion of the universe and the dilatation of distances. If $R(t)$ increases with time, then space 'dilates' and the galaxies are receding from each other not because of their own velocities but because space is in a 'state of expansion'.

An essential question in cosmology is therefore to determine the behavior of the scale factor $R(t)$, which we obtain by resolving the equations of general relativity, which relate this quantity to the energy-matter content.

From the observational point of view, one way of testing the curvature and the expansion consists in measuring the distances of ever more remote celestial bodies. In a curved space, several notions of distance are involved. If we are concentrating on the brightness of objects, then distance-luminosity D_L is important.

In 'ordinary' Euclidean space, the brightness l of an object of luminosity L varies as the inverse square of its distance, i.e.

$$l = L/4\pi D^2_L$$

The same definition is conserved in cosmology.

In a universe where space is curved and which contains various cosmological fluids, the quantity D_L becomes a complex expression as a function of redshift z. For the more advanced reader, we give the expression

$$D_L = \frac{(1+z)c}{H_0 \sqrt{|\Omega_k|}} S\left\{ \sqrt{|\Omega_k|} \int_0^z [\Omega_k (1+z')^2 + \Omega_m (1+z')^3 + \Omega_\Lambda]^{-1/2}\, dz' \right\}$$

where H_0 is the Hubble constant and $S(x)$ is a function whose expression depends on the curvature: $S(x) = \sin(x)$, x, or $\sinh(x)$ according to whether the universe is closed, flat or open.

In the expression of D_L appears the curvature term Ω_k, as well as the density parameters Ω_m and Ω_Λ, corresponding to the different contributions of the cosmological fluids (matter, 'dark energy/cosmological constant'):

$$\Omega_k = -\frac{kc^2}{R^2 H_0^2},\ \Omega_\Lambda = \frac{\Lambda c^2}{3H_0^2},\ \Omega_m = \frac{8\pi G}{3H_0^2}\rho_m \text{ with } \Omega_k = 1 - \Omega_m - \Omega_\Lambda.$$

Radiation density does not appear because, as a result of the expansion, it has been negligible since a time close to the era of recombination. In the opposite case, we would have to add to the other contributions a term $\Omega_r\,(1+z)^4$.

11. The Hubble Diagram

In astrophysics we use notions of apparent magnitude m and absolute magnitude M, related to the distance D_L (in Mpc) of an object, *via* redshift z, such that:

$$m(z) = 5 \log D_L\,(z, H_0, \Omega_m, \Omega_\Lambda) + M$$

For a given family of standard candles, absolute magnitude M is therefore known (for example, from a measurement involving the local universe, which is 'independent' of cosmology).

Measuring m as a function of z (the 'Hubble Diagram') therefore allows us to constrain Ω_m, Ω_Λ and determine the model of the universe.

In practice, we also employ the magnitude difference $\Delta\,(m - M)$ (Figure A10) in order better to visualize the differences between observations and cosmological models.

Table of Constants

Physical constants (MKSA)

speed of light	c	2.99792458×10^8	$m\ s^{-1}$
Gravitational constant	G	6.67×10^{-11}	$N\ m^2\ kg^{-2}$
Planck's constant	h	6.62×10^{-34}	J s
Boltzmann constant	k	1.38×10^{-23}	$J\ K^{-1}$
Mass of electron	m_e	9.11×10^{-31}	kg
Mass of proton	m_p	1.67×10^{-27}	kg
Avagadro's number	N_A	6.02×10^{-23}	mol^{-1}
Electron charge	e	1.60×10^{-19}	C
Vacuum permittivity	ε_0	8.854187×10^{-12}	
Fine structure constant	α	1/137 ($\alpha = e^2/(2\varepsilon_0 hc)$)	
Stefan's constant	σ	5.67×10^{-8}	$W\ m^{-2}\ K^{-4} (2\pi^5 k^4/(15h^3c^2))$

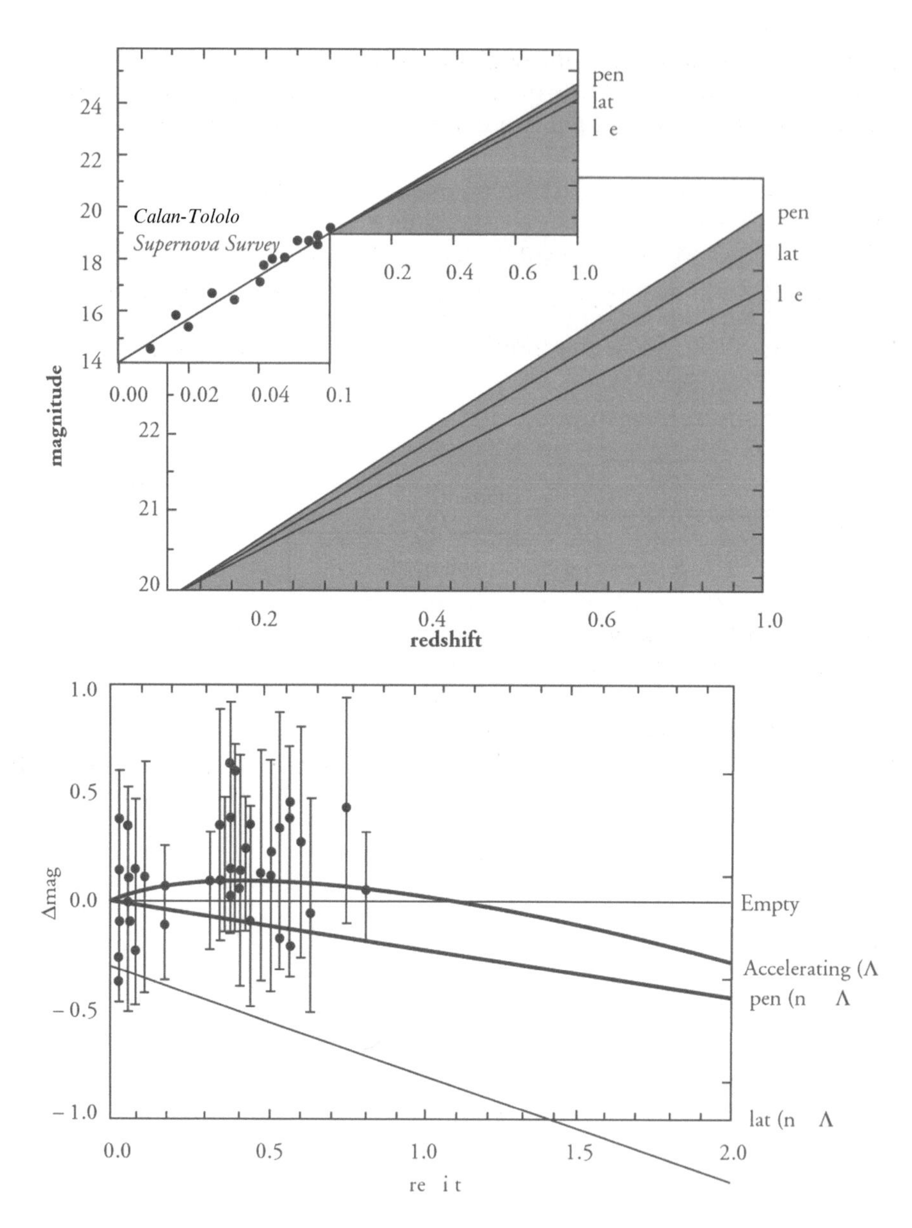

Figure A10. Above, the 'classic' Hubble diagram (magnitude-redshift) showing the expected evolution in the magnitudes of supernovae (distance-luminosity) for different models of the universe (open, flat, closed). Nearby Type Ia supernovae shown are from the Calan-Tololo Survey ($z < 0.1$). Below is introduced the quantity $\Delta(m - M)$, and the differences between the models are more clearly seen. The model of the empty universe $\Omega_{tot} = 0$ is shown for reference. The dots (with their error bars) between $z \sim 0.3$ and $z \sim 0.8$ are observations obtained by the *Supernova Cosmology Project* (SCP). These observations have revealed the acceleration of the expansion.

Units in particle physics and nuclear physics

High-energy physicists traditionally use units of time, dimension, mass, temperature etc. expressed as a function of the base unit eV (the electronvolt) and its multiples.

This system of units is obtained by allotting a value of unity (h = c = k = 1) to the fundamental constants.

In terms of *dimensional equations*, the result is that:

$$\textbf{[Energy]} = \textbf{[Mass]} = \textbf{[Temperature]} = \textbf{[Length]}^{-1} = \textbf{[Time]}^{-1}$$

Below is the rule for the conversion of these units to MKSA units, knowing that:

1 eV = 1.60×10^{-19} Joules.

	Temperature	Mass
1 eV =>	11600 K	1.78×10^{-30} kg

with $10^{-7}\ T(\mathrm{K}) \sim kT\ (\mathrm{keV})$.

Multiples: 1 keV = 10^3 eV; 1 MeV = 10^6 eV; 1 GeV = 10^9 eV.

Quantities in astronomy

$L_\odot$	Luminosity of the Sun:	3.86×10^{26} W
$M_\odot$	Mass of the Sun:	1.99×10^{30} kg
$R_\odot$	Radius of the Sun:	6.96×10^{8} m
pc	Parsec:	3.09×10^{16} m (1 Mpc = 10^6 pc, 1 Gpc = 10^9 pc)

Biblio-web

Chapter 1
Tycho Brahe: http://csep10.phys.utk.edu/astr161/lect/history/brahe.html
Chandra supernova remnant catalog:
http://hea-www.harvard.edu/ChandraSNR/gallery_gal.html
Nuclear Test Ban Treaty: http://www.ctbto.org/

Further information on **the 'local' or 'remote' nature of GRBs**: http://antwrp.gsfc.nasa.gov/diamond_jubilee/debate_1995.html

Chapter 2
Further information on **the debate concerning the determination of H_0**: http://antwrp.gsfc.nasa.gov/diamond_jubilee/debate_1996.html

Surveying the cosmos: Sloan Digital Sky Survey website:
http://www.sdss.org/
Surveying the cosmos: **The 2dF Galaxy Redshift Survey** website:
http://www.mso.anu.edu.au/2dFGRS
Space-time and the expansion of the universe:
http://rst.gsfc.nasa.gov/Sect20/A8.html
Useful **history of the universe** references:
http://astro.berkeley.edu/~jcohn/chaut/history_refs.html

Chapter 3
WMAP mission website: http://map.gsfc.nasa.gov/
Ned Wright's cosmology tutorial:
http://www.astro.ucla.edu/~wright/cosmo_01.htm
Chandra The story of stellar evolution:
http://chandra.harvard.edu/edu/formal/stellar_ev/story/

Chapter 4
Supernovae explained:
http://imagine.gsfc.nasa.gov/docs/science/know_l2/supernovae.html
Supernova classification:
http://www.jca.umbc.edu/~george/html/courses/2002_phys316/lect12/lect12_sn_basics.html

Chapter 5
An introduction to gamma-ray bursts:
http://imagine.gsfc.nasa.gov/docs/science/know_l1/bursts.html

Useful links to **GRB catalog and GRB afterglow pages:**
http://www.mpe.mpg.de/~jcg/grblink.html
Swift Satellite website:
http://heasarc.gsfc.nasa.gov/docs/swift/swiftsc.html
Gamma-ray bursts recorded by the Swift satellite:
http://heasarc.gsfc.nasa.gov/docs/swift/bursts/index.html
Binary systems of two compact objects:
http://wwwlapp.in2p3.fr/virgo/gwf.html

Chapter 6
Discovery of Cepheids and the period-luminosity relationship:
http://www.astro.livjm.ac.uk/courses/one/NOTES/Garry%20Pilkington/cepinp1.htm
The Hubble constant: the 'short-scale' and 'long-scale' controversy:
http://cfa-www.harvard.edu/~huchra/hubble/
CFHT website:
http://www.cfht.hawaii.edu/SNLS/

Chapter 7
Introduction to active galaxies and quasars:
http://imagine.gsfc.nasa.gov/docs/science/know_l1/active_galaxies.html
Distant quasar studies:
http://www.journals.uchicago.edu/ApJ/journal/issues/ApJL/v627n2/19569/19569.web.pdf
Introduction to the Lyman-alpha forest:
http://www.astro.ucla.edu/~wright/Lyman-alpha-forest.html
James Webb Space Telescope website:
http://www.jwst.nasa.gov/about.html

Chapter 8
NASA's quest for dark energy:
http://universe.nasa.gov/science/QuestForDarkEnergy.pdf
What is a cosmological constant:
http://map.gsfc.nasa.gov/universe/uni_accel.html
European Planck satellite website:
http://www.esa.int/esaSC/120398_index_0_m.html
SNAP satellite website:
http://snap.lbl.gov/
Thirty-Meter Telescope website:
http://www.tmt.org/
Extremely Large Telescopes:
http://www.oamp.fr/elt-insu/autres_liens.htm
'Quintessence' models:
http://media4.obspm.fr/public/AMC/bb/big-bang/energie-noire/bb-quintessence/index.html

Index

Printing: Mercedes-Druck, Berlin
Binding: Stein+Lehmann, Berlin